ALBEMARLE STREET

This image of some fifty men of science, including chemists, engineers, botanists, surgeons, geologists, and other natural philosophers, was produced by William Walker and George Zobel. It was first shown in the Library of the Royal Institution, where it was displayed for more than two centuries. The following are present in this assembly:

Sir Joseph Banks	Sir Benjamin Thompson	Charles Tennant
Matthew Boulton	Sir Francis Ronalds	Richard Trevithick
Robert Brown	Thomas Telford	James Watt
Sir Marc Isambard Brunel	Nevil Masikelyne	William Hyde Wollaston
Henry Cavendish	John Playfair	Thomas Young
John Dalton	Charles Hatchett	Edward Jenner
Sir Humphry Davy	William Henry	
Davies Gilbert	Sir William Herschel	

Albemarle Street

Portraits, Personalities, and Presentations at the Royal Institution

John Meurig Thomas

OXFORD

UNIVERSITY PRESS

OXFORD
UNIVERSITY PRESS

Great Clarendon Street, Oxford, OX2 6DP,
United Kingdom

Oxford University Press is a department of the University of Oxford.
It furthers the University's objective of excellence in research, scholarship,
and education by publishing worldwide. Oxford is a registered trade mark of
Oxford University Press in the UK and in certain other countries

First Edition published in 2021

Impression: 1

Published in the United States of America by Oxford University Press
198 Madison Avenue, New York, NY 10016, United States of America

British Library Cataloguing in Publication Data
Data available

Library of Congress Control Number: 2021934822

ISBN 978–0–19–289800–5

DOI: 10.1093/oso/9780192898005.001.0001

Printed and bound by
CPI Group (UK) Ltd, Croydon, CR0 4YY

*Dedicated to the memories
of
Max Perutz, Sir Ralph Kohn, and Oliver Sacks,
all of whom appreciated the significance and uniqueness of the Royal Institution*

Foreword

by

Martin Rees

It's a privilege to welcome this beautiful book, which offers a fascinating perspective on the Royal Institution (the RI), which has played a unique role in science and culture for two centuries. The book offers a perspective into many of the personalities who revolutionized our understanding of the natural world, and the RI's role in the gestation of some great scientific advances. Its author, Sir John Meurig Thomas, was able to bring unmatched credentials to the task – being an eminent researcher and brilliant expositor, who served as the RI's director during 1986-91.

In the early 19th century there was no split between 'two cultures', but a boisterous intermingling of scientists, literati and explorers. The RI was a focus of such interactions, thanks to the inspirational leadership of Humphrey Davy and Michael Faraday. These two great figures get extended coverage in the book – as does the hyper-talented but roguish adventurer Count Rumford, who donated the funds to create the RI which occupies a fine building in central London.

Rumford envisaged its mission as research, but also as the dissemination of scientific understanding among the wider population. The latter was achieved by the weekly 'discourses', involving lectures and demonstrations delivered – as they still are today – in the RI's splendid lecture theatre.

This book is a treasury of fascinating insight into the most celebrated figures who spoke at the RI, and those who made discoveries in its laboratory. But it offers, as well, 'cameos' of other speakers whose lives are less familiar. For instance, I certainly didn't know that the first fuel cell was created in the 1840s by the Welsh physicist W.T. Grove, nor that P.M. Roget, famous for his Thesaurus, was also a physiologist.

At the end of the 19th century the great Lord Kelvin famously claimed that physics was essentially 'done': all that was left was to refine the values of a few 'physical constants', and fill in some details. But Kelvin had a constricted view of the scope of science: he never conceived that we'd one day interpret the properties of materials—and even living things – in terms of their molecular structure; nor that physicists would probe deep into the nuclei of atoms (revealing an unenvisioned new source of energy); nor develop new insights into gravity, space and time. But in the 20th century, of course, all these breakthroughs happened. The scope and scale of global science widened hugely, but the RI laboratory, thanks to a roll-call of eminent directors, retained its special prestige. The RI became a forum for an even wider galaxy of speakers, with coverage spreading to fields – archaeology (especially Egyptology), space science and geophysics. And the famous Christmas lectures enthused generations of young people, especially when they reached a wider audience by being televised.

The discourses were 'performances' – daunting even to well-practiced speakers. The book is enriched by the reminiscence Thomas offers of the recent decades when he hosted visits of speakers he knew personally. He quotes extensively from many lecturers who achieved genuine eloquence in their writing and oratory. And it's good to read (pp. 201–205) from Thomas's own inaugural lecture on 'The Poetry of Science' which sustains the tradition of Davy and Faraday.

Sadly, John Meurig Thomas didn't survive to see this book in print: he died, just after its completion, in November 2020. It was, for him, a 'labour of love' to celebrate an institution he'd admired from childhood and which he'd served so well. It displays his own qualities as a polymath: he is fascinated by the personalities of scientists, and relished the 'poetry' of their work. His book is adorned with wonderful historic images of the dramatis personae spanning two centuries – among whom he will deserve a high place in future histories. It is, like the best RI lectures, insightful, eloquent and entertaining – fitting memorial to him and part of his enviable scientific legacy.

Preface

This is a personal selection of the remarkable personalities and achievements of some of the extraordinary individuals who, during the nineteenth and twentieth centuries, worked or lectured at an address in Albemarle Street in Mayfair, central London. Although most of them were among the leading natural philosophers of their day, and many were responsible for revolutionizing several facets of science and technology, others were equally renowned for their general cultural contributions to a wide range of topics encompassing the arts, literature, the humanities, drama, anthropology, Egyptology, medicine, music, poetry, politics, and religion.

Having lived for five years, and worked for twenty, at Albemarle Street, I never cease to be excited by the thought of all the great men and women who have occupied or visited number 21, which was once my home, and where my science was carried out between 1986–2006. For reasons I describe in this book, I doubt whether there is any street in London, or any other city in the United Kingdom, that has hosted such a galaxy of stars who have so profoundly influenced human life and affairs. Moreover, more famous scientists have lived at number 21 than in any other house in the world.

After its owner was murdered, number 21 Albemarle Street was sold and converted in 1799 to a laboratory, a lecture theatre, display centres, workshops, and a library, and became the Royal Institution of Great Britain. This body was granted its Royal Charter by King George III in January 1800, thanks largely to the efforts of its founder, the picaresque, American-born, Count Rumford of the Holy Roman Empire, assisted by others, such as the abolitionist and philanthropist William Wilberforce, the reclusive and taciturn Henry Cavendish, known in his time as the 'richest of the wise, and the wisest of the rich', and the powerful President of the Royal Society, Sir Joseph Banks.

The Royal Institution (henceforth RI) is the place where the first British recipient of the Nobel Prize, Lord Rayleigh, worked and gave regular popular science lectures in the period 1895–1907. It was also where Kathleen Yardley (later Dame Kathleen Lonsdale), the militant pacifist tenth daughter of an Irish postmaster, and the first woman to be elected Fellow of the Royal Society, worked for more than twenty years in the first half of the twentieth century under the aegis of Sir William Henry Bragg, another Nobel Prizewinner, and a facilitator of modern molecular biology, which had its seminal roots in Albemarle Street.

Subsequently, fifteen other Nobel Prizewinners have been associated with the RI, either as permanent or visiting scientists. But, as well as being a premier centre of scientific research in both the nineteenth and twentieth centuries, it is also (and remains) one of the most famous venues in the world for the popularization of science and the dissemination of numerous general cultural topics. It is also the place where two of the greatest physical scientists in history, Sir Humphry Davy and Michael Faraday, worked and popularized science. Hundreds of eminent authorities from the UK and abroad, in widely different fields, have also presented their work there, such as the crime writer Dorothy L. Sayers, Ernest Rutherford, Guielmo Marconi, Howard Carter, Edwin Hubble, and the US anthropologist Margaret Mead.

Davy, who was a friend of the poets William Wordsworth, Samuel Taylor Coleridge, Robert Southey, and Lord Byron, was one of the first really successful popularizers of science. He invented the influential, life-saving miner's safety lamp (for which he was rewarded by the Czar of Russia), and, for other work, presented with a gold medal by the Emperor Napoleon (when France and the UK were at war). Also, his exceptionally equipped laboratory in Albemarle Street in his day outclassed all of the scientific laboratories in the colleges of the universities of Oxford and Cambridge.

Davy's protégé, Michael Faraday, is acknowledged as one of our greatest scientists. Ernest Rutherford, a Visiting Professor at the RI, referred to Faraday, a former bookbinder, as _'the greatest discoverer ever'_. Faraday also became a brilliant popularizer of science, to children and adult lay audiences especially. His book, based on his 1860–61 Christmas Lectures for the schoolchildren of London, _'The Chemical History of a Candle'_, has been translated into more than a dozen languages. (There are currently five distinct translations of it in Japan, and schoolchildren in that country are encouraged to read it during their summer vacations.)

It was in 1826 that Faraday, as Director of the RI, initiated the practice of convening Friday Evening Discourses, at which renowned experts in all manner of disciplines and professions are invited to present the essence of their specialized topics to lay audiences in non-technical terms. He had inaugurated the Christmas Lecture Series in 1825, and he gave them nineteen times. Nowadays, the RI Christmas Lectures are televised in the UK, and since 1990 have been held in other countries, notably Japan and Singapore.

Faraday was a great favourite of Prince Albert. He also mingled with artists such as John Constable and William Brockenden, and writers such as Charles Dickens. His extraordinary succession of experimental discoveries and inventions (including the dynamo and the forerunner of the electric motor)—summarized in this book, where part of his work is compared with that of the great US polymath Benjamin Franklin, whom he greatly admired—is complemented by his contributions to theory. Albert Einstein was of the opinion that Faraday, along with James Clerk Maxwell (who was a Visiting Professor at the RI), was responsible for the greatest change in our understanding of the physical world since Sir Isaac Newton.

It is the dazzling quality of the series of the Discourses introduced by Faraday that has been partly responsible for attracting national and international celebrities to 'perform' at 21 Albemarle Street. An important subsidiary factor is the reputation that the laboratories at the RI have for being the locus of numerous seminal scientific discoveries. In the words of an early President of the US National Academy of Sciences (1868–79), Joseph Henry:

'I have always looked upon the Royal Institution as a model establishment doing honour to England, and producing an immense effect upon the world. More light has issued from that establishment in proportion to its means, than perhaps from any other on the face of the earth.'

In 2005, the then-greatest living theoretical physicist, Richard Feynman of the Californian Institute of Technology, wrote that Faraday's discovery in the 1830s, when he found out that matter and electricity, two apparently different things, were different aspects of the same thing, was *'one of the most dramatic moments in the history of science'.*[1]

Having served as the Director of the RI and of the Davy—Faraday Research Laboratory (DFRL) (occupying number 20 Albemarle Street), which was established in 1896 by the munificence of Ludwig Mond, the German-born industrialist and philanthropist, I have long felt the desire to present a palatable account of a collection of vignettes dealing with the unique legacy of this house in Albemarle Street. Not only are some of the colourful and gifted individuals from foreign lands and from the UK who have performed at the RI worthy of description, there are a host of other topics associated with the RI that merit recollection and exposure.

Much has been written already about the RI—see the early work of Martin[2] and the charming unofficial history composed by Sir William Henry Bragg's daughter, Gwendy Caroe, who lived in number 21 for nine years.[3] An elegant biography of Faraday's successor, John Tyndall, who discovered the greenhouse effect at the RI, has recently been published by Dr Roland Jackson.[4] In 2012, Sir John Rowlinson, Head of Physical Chemistry at the University of Oxford, and an ardent supporter of the RI, produced an outstanding biography of Sir James Dewar.[5] In 2006, a splendid *'Life and Scientific Legacy of George Porter'*,[6] dealing with much of George Porter's influence on the RI, was published by World Scientific Publishers. Dr John Jenkin has done great justice to the ways both Sir William Henry and his son Sir William Lawrence Bragg influenced the RI,[7] and his selection of tributes, plus Sir David Phillips' magnificent *'Biographical Memoir'* of Sir William Lawrence Bragg, was published in 1990.[8] There have also been memoirs of Sir Eric Rideal[9] and of Professor Edward Andrade,[10] two other previous Directors of the RI. But there is a place for another modern account, laced with some personal reminiscences. This monograph covers some selective aspects of the RI's existence and remarkable activities up until 1999. It is almost paradoxical that so many of the origins of modern pure and applied science, including molecular biology, with its huge impact on modern medicine, have emerged from work carried out in modest basement laboratories of converted houses (numbers 20 and 21), in a road adjacent to Lower Bond Street, that region of extravagance and sophistication in the centre of fashionable London. The RI has become known throughout the world as a scientific and cultural centre, in which our age-long search for truth, and contemplation of humankind's existence in the universe, have been examined. The RI endeavours to maintain a devotion to the ideal of its founder, that of the pursuit of knowledge for the use and benefit of humanity.

John Meurig Thomas
September 2020

1. R. Feynman, *'Thoughts of a Citizen Scientist'*, Basic Books, New York, **2005**, pp. 14–15.

2. Thomas Martin, *'The Royal Institution'*, Royal Institution, London, **1961**.

3. Gwendy Caroe, *'An Informal History of the Royal Institution'*, John Murray, London, **1985**.

4. Roland Jackson, *'The Ascent of John Tyndall'*, Oxford University Press, Oxford, **2018**.

5. J. S. Rowlinson, *'Sir James Dewar, 1842–1923. A Ruthless Chemist'*, Ashgate, Aldershot, **2012**.

6. D. Phillips and J. Barker, *'The Life and Scientific Legacy of George Porter'*, World Scientific Publishers, Singapore, **2006**.

7. J. Jenkin, *'William and Lawrence Bragg, Father and Son: The Most Extraordinary Collaboration in Science'*, Oxford University Press, New York, **2008**.

8. J. M. Thomas and D. C. Phillips, *'The Legacy of Sir Lawrence Bragg: Selections and Reflections'*, Science Reviews Ltd, London, **1990**.

9. D. D. Eley, *Biographical Memoirs of Fellows of the Royal Society*, **1976**, *22*, 381–413.

10. A. Cottrell, *Biographical Memoirs of Fellows of the Royal Society*, **1972**, *18*, 1–20.

Acknowledgements

Ever since I ceased to work at the Royal Institution in 2006, many of my friends, former colleagues there, and members have urged me to write a semi-autobiographical account of its uniqueness. Foremost among those who kept reminding me to undertake this task have been my daughters, Lisa and Naomi, and ex-colleagues, notably Professor Richard Catlow, now Foreign Secretary of the Royal Society, and Professor David Phillips, former Head of Chemistry at Imperial College, London.

Some of the contents of this collection of essays I have published previously in the *Proceedings of the American Philosophical Society*, Philadelphia, and I am grateful to that body for allowing me to repeat a few of the observations I made earlier on Rumford, Faraday, and Davy.

With the arrival of the 'lockdown' caused by the Covid-19 pandemic, I was able to focus almost exclusively on the writing of this book. Thanks to several individuals cited below, I managed to assemble a coherent account of the remarkable nature of the Royal Institution. I am greatly indebted to the following for reading a draft chapter or two and proffering helpful advice: Professors Graham Richards, Oxford; Jack Dunitz, Zurich; Dudley Herschbach, Harvard; Ted Davis, Leicester and Cambridge; Peter Edwards and Peter Dobson, Oxford; Chris Calladine and Sir Colin Humphreys, Cambridge; Dr Richard Henderson, the Laboratory of Molecular Biology, Cambridge; Dr Gari Owen, Kent; Dr D. J. Johnstone, Cambridge; the Astronomer Royal, Martin Rees, who critically perused two of my draft chapters; Dr Michelle Kohn, who read Chapter 1; and Professors Terry Jones, Adrian Dixon, and J. D. Pickard for making constructive comments on Chapter 8, and Professor Gadala-Maria of South Carolina who helped me locate obscure sources.

Providing invaluable support to me in the retrieval of images and information during the lockdown were Professor Kenneth Harris, Cardiff, and Dr D. J. Johnson, Cambridge. In addition, I received considerable help from the archivists at the Royal Institution: Professor Frank James (now of University College London) and Dr Charlotte New. Professor James especially helped me identify the source of some arcane information. Annette Faux, the Laboratory of Molecular Biology, Cambridge, as well as Sir Allan Fersht and Professor Malcolm Longair, the Cavendish Laboratory, and the staff of the Library of the Royal Society, also gave me invaluable help in retrieving illustrations. I am also grateful to Sonke Adlung of Oxford University Press for his constant guidance.

I am deeply indebted to the present Lord Rayleigh, and Lady Rayleigh, for providing many of the photographs in Chapter 5 and for general information pertaining to John William Strutt.

Valuable administrative help was provided by Liz Wake and Alison Pritchard Jones. My personal assistant, Linda Webb, as usual, was exceptional in all her work, not only translating my often indecipherable handwriting to immaculate typescript, but in helping me locate copyright owners and special illustrations. I am also grateful for the great practical help given to me by Ana Talaban-Bailey and her colleagues, Wayne Skelton-Hough and Liam Clarke, of the Department of Materials Science, Cambridge. My daughter, Lisa, gave me invaluable literary guidance: she read every word of my text and corrected several infelicities.

I am deeply indebted to Martin Rees for composing the Foreword. Ever since we first met in 1978, when I joined him as a Fellow of King's College, Cambridge, I have been extraordinarily impressed and inspired by his wide-ranging knowledge, his insights, and his exceptional expository skills.

Every day at breakfast for the last six months, my wife, Jehane, and I discussed the contents of the book. I cannot thank her enough for all the inspiring support.

To all of these kind folk, I renew my deep indebtedness and profound thanks. The responsibility for any errors and faults in this book rests with me.

Contents

1

Setting the Scene

1.1 Introduction

Less than a mile from the statue of Eros, and also from St James' Palace, lies Albemarle Street, which runs parallel to Lower Bond Street and perpendicular to Piccadilly. This is my favourite place in London. It does not compete in antiquity with the Old Kent Road or Oxford Street, for these were built by the Romans, and it does not rival in elegance the majestic splendour of Carlton House Terrace or Regent Street. But, in my opinion, it outclasses these and all other streets in the city because of its uniquely rich cultural and intellectual associations, and because of the vast range and number of world-renowned individuals who have visited it, worked at, or lectured there. Specifically, number 21, the Royal Institution (RI).

Often, as I enter it from Piccadilly and walk in the direction of Brown's Hotel (frequented by Oscar Wilde, Robert Louis Stevenson, and J. M. Barrie, and which also hosted Theodore Roosevelt, Napoleon III, and Rudyard Kipling) and pass house number 50, I am tempted to begin reciting silently the magical first verse of one of the poems that excited me as a child: Samuel Taylor Coleridge's *'Kubla Khan'*:

> *In Xanadu did Kubla Khan*
> *A stately pleasure-dome decree:*
> *Where Alph, the sacred river, ran*
> *Through caverns measureless to man*
> *Down to a sunless sea.*

It was John Murray, in number 50, who published Coleridge's collection of poems that included *'Kubla Khan'* in 1816. Nine years earlier, in the RI, Coleridge gave a series of disjointed lectures on *'The Principles of Poetry'*, allegedly under the influence of opium, to which he was addicted. It was his friend, the future President of the Royal Society Sir Humphry Davy, who was effectively in charge at that time in number 21, and he paved the way for Coleridge to present his set of somewhat disorganised and disjointed lectures.[1] Among other things, he held forth on the topic of psychosomatic illnesses, one that interested Coleridge greatly.[2]

Albemarle Street: Portraits, Personalities, and Presentations at the Royal Institution. John Meurig Thomas,
Oxford University Press. © Sir John Meurig Thomas 2021.
DOI: 10.1093/oso/9780192898005.003.0001

John Murray also published all the poetic works of Lord Byron, including the extraordinarily successful *'Child Harold's Pilgrimage'*, the first printing of which sold out in five days. In May 1824, John Murray's premises witnessed one of the most notorious acts in the annals of English literature when the owner destroyed the manuscript of Byron's memoirs because it was felt that their scandalous details would destroy Byron's reputation.

From 1813 to 1818, John Murray also published (anonymously) Jane Austen's *'Pride and Prejudice'* and her other five major novels, which have all remained in print since. Sir Arthur Conan Doyle (of Sherlock Holmes fame) and the American author Herman Melville also published some of their works in Albemarle Street. Indeed, Melville's *'Typee: A Peep at Polynesian Life'* (1846) and *'Omoo: A Narrative of Adventure in the South Seas'* (1847) caused quite a stir when they appeared.

At about the time of publication of these two books by Melville, Peter Mark Roget, the physician, natural theologian, and lexicographer, working as a physiologist at number 21 Albemarle Street, was engaged in the final stages of his *'Thesaurus of English Words and Phrases'*. Roget had commenced work on this venture some thirty years earlier, partly as a means of keeping at bay his battles with depression (see Chapter 10).

On 24 November 1859, one of the most influential books ever written, Charles Darwin's *'On the Origin of Species by Means of Natural Selection'*, was published by John Murray. Number 21 Albemarle Street was later to be the scene where many lively discussions on the validity and consequences of the theory of evolution were held.

1.2   Some Eminent Visitors to Number 21 Albemarle Street in the Late Nineteenth and Early Twentieth Centuries

In May 1889, for example, the extraordinarily charismatic Russian chemist, encyclopaedist, and inventor (also a close friend of his St Petersburg contemporary, the composer and fellow chemist Aleksander Borodin) Dmitri Ivanovich Mendeleev came to lecture at the RI. Mendeleev was the youngest of fourteen children born in Tobolsk in Siberia. His mother recognized his precocious intelligence, and when he was fourteen, eager to educate him properly, walked thousands of miles from Siberia with him to St Petersburg, where he got a grant to train as a teacher. Even as a student, Mendeleev showed not only an insatiable curiosity, but a hunger for organizing principles of all kinds. This quality reached its climax in 1869 when, after twenty years of deep cogitation and lucubration, he produced his brilliant periodic table of the elements.

Every scientist I have known who has ever contemplated the information contained in Mendeleev's periodic table confesses to being bowled over by it. The novelist C. P. Snow's reaction was as follows:

'For the first time I saw a medley of haphazard facts fall into line and order. All the jumbles and recipes and hotchpotch of the inorganic chemistry of my boyhood seemed to fit themselves into the scheme before my eyes—as though one were standing beside a jungle and it suddenly transformed itself into a Dutch garden.'

Oliver Sacks, in his '*Uncle Tungsten*',[3] tells us that, after seeing Mendeleev's periodic table in the Science Museum as a youngster in 1945:

'I could scarcely sleep for excitement the night after seeing the periodic table—it seemed to me, an incredible achievement to have brought the whole, vast and seemingly chaotic universe of chemistry to an all-embracing order…to have perceived an overall organisation, a super-arching principle uniting and relating all the elements, had a quality of the miraculous, of genius. And this gave me, for the first time, a sense of the transcendent power of the human mind, and the fact that it might be equipped to discover or decipher the deepest secrets of nature, to read the mind of God.'

Mendeleev had classified the elements according to their atomic weights and valency, concepts that are clarified later in this book. Oliver Sacks, on seeing the photograph of Mendeleev next to the periodic table in the museum, said that

Figure 1.1 *(a) Dmitri Ivanovich Mendeleev, who gave a Friday Evening Discourse at the Royal Institution (RI) in 1889 and again a few years later. The stamp reproduction here, created by Evgeni Egorov, was printed in Russia to commemorate both the 175th anniversary of Mendeleev's birth and the 140th anniversary of the publication of his periodic table of the elements. (b) Dmitri Ivanovich Mendeleev:* 'The discovery and isolation of sodium and potassium was one of the greatest discoveries in science', 'The Principles of Chemistry', *Longmans, Green and Co.,* **1905**.

(Courtesy Erling Norrby and Ulf Lagerkvist)

the Russian looked like a cross between Fagin and Svengali. But, Mendeleev was the gentlest of human beings, and for his imaginative science he was awarded the Davy Medal of the Royal Society in 1892 and, in 1906, that Society's premier award, the Copley Medal.

Madame Curie, who accompanied her husband, Pierre, in June 1903 on a visit to Albemarle Street—where he lectured on '*Radium*' at number 21—was,

(a)

(b)

Figure 1.2 *(a) Pierre and Marie Curie (provided by Professor Michel Che, Paris) (b) Sir James Dewar, inventor of the thermos flask (Dewar vessel), at his desk at the Davy—Faraday Research Laboratory.*

(Courtesy JET Photographic and the Master and Fellows of Peterhouse, Cambridge)

along with her husband, awarded the Nobel Prize in Physics four months later for their pioneering work on radioactivity in the late 1890s and early years of the twentieth century. The Curies often visited Sir James Dewar (the inventor of the thermos flask), who lived and worked at the RI from 1877 to 1923 as both its Director and as Co-Director with Lord Rayleigh of the Davy–Faraday Research Laboratory (DFRL).

In the 1890s also, John Ruskin, the indefatigable and brilliant prophet of the Victorian age, spoke at the RI (to an audience of 1,144) on *'Verona'*, and the novelist, journalist, scientific visionary, and encyclopaedist H. G. Wells too lectured there on the *'Discovery of the Future'* shortly thereafter. The roll call of distinguished visitors who spoke at the RI in days of yore is exceptionally interesting. The following are noteworthy:

- John Constable, who gave a series of lectures on the history of landscape painting at the RI in 1836, to which Sir Ernst Gombrich, the art historian, made reference in his Discourse there in 1978
- Fox Talbot, the founding father of photography
- Charles Babbage, the originator of the first programmable computer
- Ada, the Countess of Lovelace, pioneer computer scientist, Babbage's friend, and Lord Byron's daughter, who became infatuated with Michael Faraday
- Guglielmo Marconi, the Italian inventor of radio
- H. Becquerel, the French discoverer of radioactivity
- Mary Somerville, geographer, mathematician, and polymath, renowned also for the Oxford college named after her
- Nikola Tesla, the Serbian-American inventor and electrical engineer
- Logie Baird, the Scottish inventor of television
- Howard Carter, discoverer of the tomb of Tut-ankh-Amun
- Ernest Rutherford, instigator of the nuclear age, known as the Newton of the atom, and regarded as the greatest experimentalist since Faraday
- William Thomson, Lord Kelvin, initiator of the Atlantic cable, after whom the absolute scale of temperature is named
- The Lords Adrian, physiologists, father and son
- Walford Davies and Hubert Parry, composers
- Yehudi Menuhin, violinist
- Arthur Bryant, historian renowned for his study of Samuel Pepys
- J. B. Priestley, novelist and broadcaster
- Jacquetta Hawkes, archaeologist and writer
- Rosalind Franklin, early crystallographer of viruses and DNA
- Erwin Schroedinger, father of quantum mechanics
- Edward Teller, described as the father of the hydrogen bomb

- David Attenborough, naturalist and broadcaster
- Marcus Sieff, businessman
- Linus Pauling, effulgent US chemist and molecular biologist
- Johnathan Miller, opera and theatre director
- E. M. Forster and Charles Morgan, writers
- Max Perutz and John Kendrew, and their junior colleagues Francis Crick and J. D. Watson, all founders of modern molecular biology
- Sydney Brenner, architect of the genetic code
- C. P. Snow, novelist
- James Jeans, astronomer
- J. J. Thomson, discoverer of the electron
- Sergey Kapitza, Soviet physicist and broadcaster

In subsequent chapters, accounts are given of many of the crucial early occupants of the RI that were responsible for conferring upon it its unique qualities.

It is important to note that Rumford, the RI's founder, early on appointed two exceptionally gifted West Country Englishmen, Thomas Young (of Milbarton, Somerset) and Humphry Davy, as well as a versatile Clerk-of-the-Works, Thomas Webster, an adventurous Scot from the Orkney Islands. Young was appointed Professor of Natural Philosophy, Davy as a lecturer in chemistry, and Webster was mainly responsible for creating one of the world's finest lecture theatres, at 21 Albemarle Street. Chapter 2 deals rather fully with the other aspects of Rumford's remarkable creations, discoveries, and innovations before he left London for Paris.

Humphry Davy (Figure 1.3), whose life, work, and legacy is described in Chapter 3, combined brilliant scientific research with outstanding presentational skills as a popularizer of science, especially chemical science and aspects of antiquity. So popular were his lectures at the RI in the early decades of the nineteenth century with the socialites, intellectuals, and aristocrats of London that they flocked in their horse-drawn carriages to his performances. In order to ease the traffic congestion, Albemarle Street became the first one-way thoroughfare in the metropolis.

Thomas Young (Figure 1.4), who occupied his professorship from 1801–03, was also a brilliant scientist, but far inferior to Davy as a lecturer. He was regarded by his contemporaries as the very greatest natural philosopher in view of the depth and range of his understanding and talents. Before leaving the RI, Young had completed work of fundamental importance in elucidating the phenomena of capillarity, elasticity, and vision. His so-called two-slit experiment[4] (see Chapter 11) turned out to be so important that it is still the subject of lively debate, and claimed by some physicists to be one of the most beautiful experiments in natural philosophy.[5]

Figure 1.3 *Sir Humphry Davy, Bart 1778–1829; Fellow of the Royal Society 1803; Secretary of the Royal Society 1807–12; President of the Royal Society 1820–27.*

(Courtesy The Royal Society)

Figure 1.4 *Portrait of Thomas Young (1773–1829).*

(Courtesy The Royal Society)

But, it was Davy who established the RI as a major centre of scientific research; the presentations that he gave and the discoveries and inventions that he made in Albemarle Street turned out to be of everlasting importance, as is described in Chapter 3.[6] However, Davy often remarked that his greatest discovery was a London-born, former bookbinder's apprentice.

Michael Faraday, who, after William T. Brande's role as Director following Davy's retirement, subsequently took over as the key investigator and lecturer in the RI, where he worked for fifty-four years. It was, as stated earlier, he who introduced the practice of holding Friday Evening Discourses in the spring, autumn, and winter months of each year, of which more will be elaborated later in Chapter 4. We simply note here that the sheer diversity of specialists that he and the managers of the RI were able to attract was remarkable: poets, painters, Assyriologists, Egyptologists, explorers, musicians, economists, chemists, cosmologists, and classicists, quite apart from experimental and theoretical scientists. These speakers enlightened and expanded the horizons of the intelligent lay members of the RI, sometimes with the aid of graphic demonstration. It was in the lecture theatre of 21 Albemarle Street in May 1861 that Faraday had arranged for James Clerk Maxwell (who later was subsequently described as the Scottish Einstein) to demonstrate the feasibility of colour photography in a Discourse entitled *'On the Theory of Three Primary Colours'*. This was a subject that Thomas Young had researched and illuminated half a century earlier in the RI.

Figure 1.5 *(a) Michael Faraday (drawing by G. Richmond, 1852). (b) James Clerk Maxwell.*
((a) Courtesy RI (b) Courtesy Cavendish Laboratory, Cambridge)

Faraday himself gave one of the earliest public demonstrations of photography at a Discourse with the cooperation of Fox Talbot. Indeed, the first recorded electrical flash photograph was taken in the lecture at the RI. In a letter addressed to Faraday, dated 15 June 1851, Fox Talbot writes: *'If a truly instantaneous photographic representation of an object has never been obtained before (as I imagine it has not), I am glad that it should have been first accomplished at the Royal Institution.'*

Faraday's successor as the supervisor at the RI was the unusual, German-educated Irishman John Tyndall, who was an exceptional expositor and a fine experimentalist. It was Tyndall who first identified the phenomenon of the greenhouse effect, now a matter of great concern in the context of climate change. Tyndall, and his successor, Sir James Dewar, as well as their successors as Directors of the RI, notably Sir William Henry (Chapter 7), not only maintained the tradition of pursuing scientific research of Nobel Prizewinning quality—the father and son, William Henry and Lawrence Bragg, shared the Nobel Prize in Physics in 1915—but they also recruited an admirable collection of speakers for Discourses and other special lectures. In this they maintained the tradition of Faraday and Dewar. Thus, in an eight-year period from 1900, the following Nobel Prizewinners delivered Friday Evening Discourses (Table 1.1):

Later Directors of the RI followed[7,8] the example of Faraday, Tyndall, and Dewar in enlisting academics from the University of Cambridge as Visiting

Table 1.1 *Nobel Prizewinners who gave Discourses in James Dewar's era as Director of the RI*

Date of Discourse	Speaker (Nationality)	Topic	Date of Nobel Prize
02/02/1900	G. Marconi (Italian)	Wireless telegraphy	1909
02/03/1900	R. Ross (English)	Malaria and mosquitoes	1902
07/03/1902	A. H. Becquerel (French)	The radioactivity of matter	1903
19/06/1903	P. Curie (French)	Radium	1903
19/02/1904	C. T. R. Wilson (Scottish)	Condensation nuclei	1927
20/05/1904	E. Rutherford (New Zealander)	The radiation and emanation of radium	1908
03/06/1904	S. Arrhenius (Swedish)	The development of the theory of electrolytic dissociation	1905
30/03/1906	P. Zeeman (Dutch)	Recent progress in magneto-optics	1902
19/04/1907	C. S. Sherrington (English)	Nerve as a mastery of muscle	1932

Figure 1.6 *Professor of Physics at the RI, Ronald King, assisted by the versatile technical assistant 'Bill' Coates, giving a schools lecture at the RI in the 1960s.*

(Courtesy Professor Ronald King)

Professors. Two very famous nineteenth-century scientists in this category were James Clerk Maxwell and the Nobel Prizewinner J. J. Thomson, who announced the discovery of the first-ever identified subatomic particle, the electron, during a Discourse at the RI. Other famous individuals who later became Visiting Professors and who gave Discourses and other special lectures to lay audiences at the RI were: Ernest (Lord) Rutherford, the farmer's son from New Zealand; Sir Brian Pippard, a successor of Rutherford's as Cavendish Professor at Cambridge; Antony Hewish, the Nobel Prizewinning co-discoverer of quasars; and Charles A. Taylor, Professor of Physics at Cardiff and one of the world's greatest authorities on the physics of music. Taylor, who was appointed Visiting Professor of Experimental Physics at the RI in 1971, when he also gave the Christmas Lectures on *'Sounds of Music: The Science of Tones and Tunes'*, gave, in all, some one-hundred-and-fifty lectures for schoolchildren on various aspects of physics, as well as eight Friday Evening Discourses at the RI. (In 1989, I invited him to give his second series of Christmas Lectures in *'Exploring Music'*. In the following year, he accompanied me to Japan, and he and I replicated one of our respective RI Christmas Lectures in an enormous hall holding several thousand children, in Tokyo, under the auspices of the British Council and the Japanese newspaper, *Yomiuri Shimbun*.) Further notables included Sir Christopher Zeeman, Professor of Mathematics at the University of Warwick, renowned for his work on catastrophe theory, who was Visiting Professor of Mathematics at the RI, and Professor Ronald King, an expert on metal physics, who occupied a chair of physics at the RI for many years and who was responsible, with Sir Lawrence Bragg, for introducing the series of regular schools' lecture-demonstrations at the RI in the mid-1950s. In addition, Professor Anne McLaren, the distinguished biologist, was Visiting Professor of Physiology for many years.

1.3   The Unique Qualities of the RI

Throughout the nineteenth and twentieth centuries, the RI, an independent body, and an educational charity, functioned as part university, part museum, part research centre, part classroom, part library, part London club, and part exhibition and broadcasting centre. The eminent author and broadcaster Sir Antony Jay (of television's *Yes, Prime Minister* fame) frequently described the lecture theatre of the RI as a perfect venue for broadcasting.

Directors and managers of the RI over the Davy—Faraday period (1801–62) and later encountered little difficulty in attracting the foremost authorities on essentially all subjects from the UK and abroad to perform there, to discriminating and welcoming audiences. This is because of the exceptionally high reputation of the prior occupants of the Directorships and Visiting Professorships of previous eras; and especially because of the aura exuded by Davy and Faraday, as well as the acknowledged distinction of their successors—Tyndall, Dewar, Rayleigh, the

Figure 1.7 *David Phillips presenting his famous Friday Evening Discourse at the RI in November 1965.*

(By kind permission of Louise Johnson)

Braggs, Ridael, Andrade, and Porter, all of whom were themselves exceptionally gifted disseminators of knowledge, as well as being innovative researchers (Figure 1.8). The scientific and almost quasi-spiritual presence of Michael Faraday at the RI confers a unique atmosphere that still pervades the whole place.

At present, however, we focus on another aspect of his impact on the propagation of human knowledge. This devolves upon his inception of the RI Christmas Lectures.

1.4 RI Christmas Lectures

In 1825, Faraday instituted the RI Christmas Lectures for the schoolchildren of London and its environs. From the outset, up to seven-hundred children attended these lectures, the first of which was given by Faraday's colleague John Millington, who was Professor of Mechanics at the RI. Up until the 1839–40 vacation, Faraday

(a)   (b)

Figure 1.8 *(a) Sir George Porter, Baron Porter of Luddenham, OM, President of the Royal Society, Fellow of the Royal Society of Edinburgh (b) Sir Peter Medawar, OM, CBE, Fellow of the Royal Society.*

(a) and (b) Wikipedia open access

Table 1.2 *Titles of the RI Christmas Lectures given by Michael Faraday in the 1850s*[8]

1851–2	'Attractive Forces'
1852–3	'Chemistry'
1853–4	'Voltaic Electricity'
1854–5	'The Chemistry of Combustion'
1855–6	'The Distinctive Properties of the Common Metals'
1856–7	'Attractive Forces'
1857–8	'Static Electricity'
1858–9	'The Metallic Properties'
1859–60	'The Various Forces of Matter and their Relations to Each Other'
1860–61	'The Chemical History of a Candle'

gave five of these sets of lectures, three on chemistry and two on electricity. In all, he gave nineteen sets of Christmas Lectures; ten of them in succession in the period 1851–61 (see Table 1.2). All these were popular.

During December 1856, Faraday delivered one of his lectures in the presence of Prince Albert and two of his sons (Figure 1.9). (A variant of this engraving (by

Figure 1.9 *Michael Faraday delivering a Christmas Lecture on 27 December 1856 in the presence of Prince Albert, the Prince of Wales, and his brother. (After Alexander Blaikley.)*
(Courtesy RI)

Figure 1.10 *Michael Faraday, Bank of England £20 note.*
(Courtesy Bank of England)

Watkins) was incorporated in the £20 note, which I persuaded the Bank of England to print (replacing the image of William Shakespeare) on the occasion of the bicentenary of the birth of Faraday in 1991.)

Faraday's enviable reputation as a lecturer and communicator to young audiences—a juvenile auditory as he used to call it—grew markedly in this period. In 1856, for example, the *London Illustrated News* wrote:

> *'For the last eight seasons Professor Faraday has undertaken this task with a modesty and a power which it is impossible to praise[9] too much. There can be no greater treat to any one fond of scientific pursuits than to attend a course of these lectures.'*

The announcement of the first series of lectures given by Faraday in 1827 is shown in Figure 1.11.

His last series (1860–61) was on a topic he had first delivered in the 1948–9 vacation, and this turned out to be quite sensational at the time, and also subsequently: *The Chemical History of a Candle*. Before commenting further on this and its subsequent enormous impact, it is noteworthy to observe that Faraday, during this period, was heavily involved in a host of other activities. He published, in 1827 alone, seventeen original papers, the titles of which are given in Table 1.3.

That year also saw the appearance of the first editions of Faraday's 646-page book—the only one he wrote—on '*Chemical Manipulation*', a monograph to which, many decades later (1931), the Nobel Laureate in Chemistry Sir Robert Robinson, the premier chemist of his age, accorded great praise.[7]

We now dwell further on Faraday's '*Candle*' lectures. Even in the course of their presentation, such was Faraday's reputation that they were being reported in the press. In the UK, *The Morning Post* (December 1860) and the *British Medical Journal*, as well as *The Times*, reported his lectures, which elicited some correspondence from readers. The versatile Sir William Crookes,[12] friend of Faraday, published, verbatim, lectures by Faraday in his journal *Chemical News* for 1861. Moreover, the *Scientific American*, in the US, also devoted two issues to each of Faradays lectures.

This series on the '*Candle*' has become a world classic. It has been translated into more than a dozen languages, including Latvian, Bulgarian, and Slovenian. The reasons for its continuing popularity are numerous. First, its lucid treatment of the combustion of a candle is still valid, and is beautifully described. Second, its illustrations are instructive, and the explosions that it describes win the hearts of young people. Third, the language is simple and attractive, and teachers of English to foreign students regard it as exemplary. Lastly, the near theistic (moral) philosophy that permeates the text appeals to teachers and students for sound social reasons. Nowhere is this more apparant that in the final paragraph of Faraday's sixth lecture:

> *'Indeed, all I can say to you at the end of these lectures (for each must come to an end at one time or other) is to express a wish that you may in your generation, be fit to compare to a*

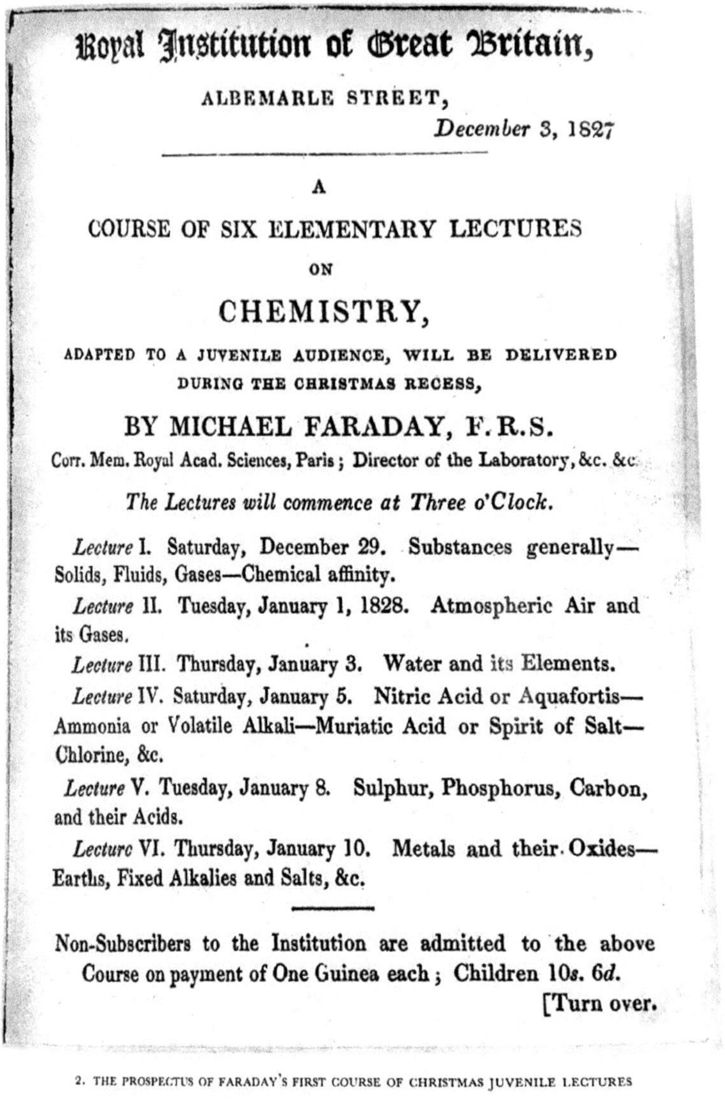

Royal Institution of Great Britain,

ALBEMARLE STREET,

December 3, 1827

A

COURSE OF SIX ELEMENTARY LECTURES

ON

CHEMISTRY,

ADAPTED TO A JUVENILE AUDIENCE, WILL BE DELIVERED
DURING THE CHRISTMAS RECESS,

BY MICHAEL FARADAY, F.R.S.

Corr. Mem. Royal Acad. Sciences, Paris; Director of the Laboratory, &c. &c.

The Lectures will commence at Three o'Clock.

Lecture I. Saturday, December 29. Substances generally—
Solids, Fluids, Gases—Chemical affinity.

Lecture II. Tuesday, January 1, 1828. Atmospheric Air and
its Gases.

Lecture III. Thursday, January 3. Water and its Elements.

Lecture IV. Saturday, January 5. Nitric Acid or Aquafortis—
Ammonia or Volatile Alkali—Muriatic Acid or Spirit of Salt—
Chlorine, &c.

Lecture V. Tuesday, January 8. Sulphur, Phosphorus, Carbon,
and their Acids.

Lecture VI. Thursday, January 10. Metals and their Oxides—
Earths, Fixed Alkalies and Salts, &c.

Non-Subscribers to the Institution are admitted to the above
Course on payment of One Guinea each; Children 10*s*. 6*d*.

[Turn over.

2. THE PROSPECTUS OF FARADAY'S FIRST COURSE OF CHRISTMAS JUVENILE LECTURES

Figure 1.11 *The Prospectus of Faraday's First Course of Christmas Juvenile Lectures, 1827.*
(Courtesy RI)

> *candle; that you may, like it, shine as lights to those about you; that, in all your actions, you*
> *may justify the beauty of the taper by making your deeds honourable and effectual in the*
> *discharge of your duty to your fellow-men.'*

In my opinion, and that of many others, the very first sentence of Lecture 1 of the
'*Candle*' series is charming and reinforces the genius of Faraday:

'I propose, in return for the honour you do us by coming to see what are our proceedings here, to bring before you, in the course of these lectures, the Chemical History of a Candle. I have taken this subject on a former occasion, and were it left of my own will, I should prefer to repeat it almost every year; so abundant is the interest that attaches itself to the subject so wonderful and to the varieties of outlet which it offers into the various departments of philosophy. There is not a law under which any part of the universe is governed which does not come into play and is touched upon in these phenomena. This is no better,

Table 1.3 *List of the Original Papers Published by Faraday in the Year he delivered his First Series of RI Christmas Lectures (1827)*

1. 'Plan of an extended and practical course of lectures…delivered in the laboratory of the RI by W. T. Brande and M. Faraday', *Quart. J. Sci.*, **1827**, *22*, 231.

2. 'On a peculiar perspective appearance of aerial light and shade', *Quart. J. Sci.*, **1827**, *22*, 81.

3. 'On the confinement of dry gases over mercury', *Quart. J. Sci.*, **1827**, *22*, 220.

4. 'A general account accompanied with experimental illustrations of the late extension of our knowledge relative to magnetism, found on the discovery of M. Arago, of the effects of metal when in motion', *Quart. J. Sci.*, **1827**, *new ser. 1*, 209.

5. 'London Institution, 1827 Prospectus of Mr Faraday's Lectures on the philosophy and practice of chemical manipulation, February 13—May 8'.

6. 'Measurements and weights of a sea-gull', *Quart. J. Sci.*, **1827**, *new ser. 1*, 244.

7. 'On the probable decomposition of certain gases compounds of carbon and hydrogen, during sudden explosion', *Quart. J. Sci.*, **1827**, *new ser. 1*, 204.

8. 'On some general points of chemical philosophy short extracts', *B. J., 1*, 353.

9. 'On the chemical action of chlorine and its compounds as disinfectants', *Quart. J. Sci.*, **1827**, 460.

10. 'On the progress and present state of the Thames Tunnel', *Quart. J. Sci.*, **1827**, 466.

11. 'Transference of heat by exchange of capacity in gas', *Quart. J. Sci.*, **1827**, 474.

12. 'Corrections in a work entitled Chemical Manipulation', *Phil. Mag.*, **1827**, *3*, 58.

13. 'Letter to Ampère dated RI September 5, 1827', *Correspondense du Grand Ampère*, Ed Lat de January, Paris, **1936**.

14. 'Letters and demonstrations in theoretical and practical chemistry commencing 9th Oct, 1827 by Brande and Faraday', *Quart. J. Sci.*, **1827**, 4.

15. 'Descriptions of two remarkable ores of copper from Cornwall by William Phillips, with analysis of the same by M. Faraday', *Phil. Mag.*, **1827**, *2*, 286.

16. 'Experiments on the nature of Labarraque's disinfecting soda liquid', *Quart. J. Sci.*, **1827**, 84.

17. 'Faraday's Chemical Manipulations', *Quart. J. Sci.*, **1827**, *new ser. 2*, 23.

18. 'A course of six elementary lectures on chemistry, adapted to a juvenile audience by Michael Faraday', *Quart. J. Sci.*, **1827**, viii (a lecture prospectus).

19. 'On the fluidity of sulphur and phosphorous at common temperatures', *Quart. J. Sci.*, **1827**, *2*, 469.

there is no more open door by which you can enter into the study of natural philosophy, than by considering the physical phenomena of a candle.'

Since 1986, I have lectured on the topic of '*The Genius of Michael Faraday*' more than three-hundred times in five continents, to young children and adults of all ages. In my experience, most members of the audiences are greatly enthused by the description given of Faraday's series of Christmas Lectures on the *'Candle'*. In addition to the science and almost spiritual flavour of the lectures, they are also transported by the opening of the preface to the lectures that are reproduced below.

'From the primitive pine torch to the paraffin candle, how wide an interval! Between them how vast a contrast! The means adopted by man to illuminate his home at night, stamp at once his position in the scale of civilisation. The fluid bitumen of the Far East, blazing in rude vessels of baked earth; the Etruscan lamp, exquisite in form, yet ill adapted to its office; the whale, seal or bear fat, filling the hut of the Esquimoux or Lap with odour rather than light; the huge wax candle on the glittering altar; the range of gas lamps in our streets—all have their stories to tell. All, if they could speak (and after their own manner they can) might warm our hearts in telling how they have ministered to man's comfort, love of home, toil and devotion!'

1.5   RI Christmas Lectures in Japan

Young and old scientists, the world over, who have read the *'Candle'* series by Faraday frequently assert how indebted they are to it. I have the feeling that in no country on earth is this indebtedness felt more than in Japan. In the first place, there are several different translations of it—see Figure 1.12, taken from a recent letter sent to me by the distinguished Japanese scientist Professor Osamu Terasaki. Note that in Japan, even very young children are offered special versions of the translation.

In the second place, many distinguished Japanese scientists frequently acknowledge publicly how indebted they are to Faraday for his *'Candle'* lectures. One of the most recent examples is Professor Akira Yoshino of the Asahe Kasei Corporation (and University of Meijo), joint-recipient of the 2019 Nobel Prize in Chemistry. He is, along with his co-recipients John B. Goodenough and M. S. Whittingham, chiefly responsible for creating affordable, re-chargeable mobile phones (Figure 1.13).

At the age of nine, Professor Yoshino's schoolteacher drew his attention to Faraday's *'Candle'* series. Professor Terasaki, my former postdoctoral research associate (PDRA), who retired from the Arrhenius Professorship of Structural Chemistry in the University of Stockholm, Sweden, a few years ago, was given a copy of the *'Candle'* series by his father when he was a schoolboy. Other distinguished Japanese scientists and technologists, mentioned in Section 1.8, testified to me their admiration of Faraday in general and their indebtedness for his *'Candle'*

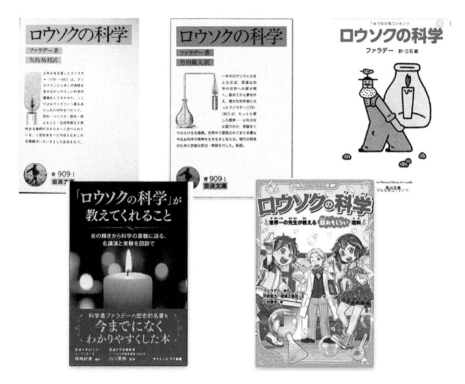

Figure 1.12 *Selection of several different translations, taken from a recent letter sent to me by the distinguished Japanese scientist (Professor O. Terasaki)*
(Courtesy Professor Osamu Terasaki)

series in particular. It is noteworthy that several authors have returned to repeat some of the experiments conducted by Faraday in his *'Candle'* series. Thus, my former PDRA Professor Wuzong Zhou,[13] now Professor of Chemistry at St Andrew's University, investigated by modern high-resolution electron microscopy the nature of the deposit when the Faraday flame from the candle was impinged upon a cold surface. Professor Zhou found evidence for the presence of Buckminsterfullerene C_{60} species in the carbon laid down when the flame of the candle impinged upon a cold metal surface.

1.5.1   More Recent RI Christmas Lectures and their Influence

Even before the RI Christmas Lectures were televised by the BBC in 1966 (a move masterminded by its then Director Sir (later Lord) George Porter), many scientists in Britain had their interests kindled by the RI Christmas

Figure 1.13 *Professor Akira Yoshino of the Asahe Kasei Corporation (and University of Meijo), joint-recipient of the 2019 Nobel Prize in Chemistry.*
(Wikipedia open access)

Lectures. Schoolchildren in the London area were prime beneficiaries, since they had the opportunity of being present at the live lectures. Several of the RI lectures were later produced as books. Thus, Sylvanus P. Thompson, who wrote a masterly account of the life of Faraday, published his 1896–7 lectures on *'Light, Visible and Invisible'*,[14] and a later Director, Sir William Henry Bragg, published both his 1923–4 and 1925–6 series on *'Concerning the Nature of Things'*[15] and *'Old Trades and New Knowledge'*,[16] respectively.

Dorothy Hodgkin (née Crowfoot), OM, the first woman to win the Nobel Prize in Chemistry, after the Curies (mother and daughter did so), did not attend any of Sir William Henry Bragg's lectures, but as a fifteen-year-old her mother (an Egyptologist) bought her the published copies of his RI Lectures, described above. In Bragg's books, the discipline of X-ray crystallography was mentioned, and this had enabled him and his son (Sir Lawrence Bragg) to 'see' atoms. Sir William Henry Bragg's elegant introduction excited the impressionable Dorothy Crowfoot beyond measure. After studying chemistry at Oxford, she took the first step towards her long-cherished ambition to see atoms by joining at Cambridge the leading young X-ray crystallographer, J. D. Bernal (who had been trained by Sir William Henry Bragg at the DFRL, where he took his Ph.D.—see Chapter 7). Together, Hodgkin/Crowfoot and Bernal were to make major advances in the

Figure 1.14 *Photograph of a characteristic discussion between Dorothy Hodgkin and J. D. Bernal taken towards the end of Bernal's life.*
(Photographer unknown)

Figure 1.15 *Sir Frank Whittle standing at the lecturer's desk before giving his RI Christmas Lecture in 1954 on the story of petroleum. His jet engines and associated apparatus are in the foreground. All the clothes and the shoes worn by the lady standing to his left were made from petroleum.*
(By kind permission of Shell International)

early studies of molecular biology. These, as well as important studies of Perutz, Kendrew, Phillips, and North in molecular biology are described in Chapter 7.

There have been several, post-World War II series of Christmas Lectures that have proved very successful. The one by the inventor of the turbo-jet aircraft, Frank Whittle (1954–5), on *'The Story of Petroleum'* (Figure 1.15), was very popular, as were those by David Attenborough (1973–4) on *'The Language of Animals'*, by the US cosmologist and astronomer Carl Sagan (1977–8) on *'The Planets'*, by George Porter (1976–7) on *'The Natural History of a Sunbeam'*, by Christopher Zeeman (1978–9) on *'Mathematics in to Pictures'*, by Charles A. Taylor (1989–90) on *'Exploring Music'*, by Malcolm S. Longair (1990–91) on *'The Origins of the Universe'*, by Simon Conway Morris (1996–6) on *'The History in our Bones'*, and by the eminent physiologist Dame Nancy Rothwell (1998–9) on *'Staying Alive'*.

A particularly memorable series was presented in 1991–2 by Richard Dawkins on *'Growing up in the Universe'*, subsequently published in the book *'River out of Eden: A Darwinian View of Life'*.[17] In Dawkins' autobiography, *'Brief Candle in the Dark: My Life in Science'*,[18] there is a fascinating account given by the author of his experience as a RI Christmas Lecturer, from the moment he was invited to do so, by me, to the intricacies of their preparation, presentation, and ultimate broadcast and publication.

1.6   Mathematics at the RI

In 1979–80, the RI Christmas Lectures were presented by Sir Christopher Zeeman of the University of Warwick on the title *'Mathematics into Pictures'*. This is what Zeeman wrote as their summary:

> *'This is the first time in 149 years that the Christmas Lectures have been devoted to mathematics. Maybe this is because it is a paradoxical subject: we are never quite sure whether it is an art or a science, whether we invent it or discover it, whether it is a man-made toy or a truth so universal that it is independent of the universe. It is one of the oldest and most splendid endeavours of mankind. Some people love it and others hate it. It can be very pure with its patterns of growth determined by its own intrinsic criteria of beauty, or it can be very applied with its patterns of growth determined by its usefulness in science.'*

In 1997–8, Professor Ian Stewart, also of the University of Warwick, gave another series on mathematics entitled *'The Magical Maze'*.

So successful were Zeeman's series that they later led the RI to establish Mathematics Masterclasses in Albemarle Street for young children. Such classes are now mounted in twenty towns and cities within the UK (thanks to the support of the Worshipful Company of Clothworkers) as well as regularly at the RI. But, in Albemarle Street in particular, as a result of an initiative that I took as Director, the Worshipful Company of Clothworkers awarded the RI a massive grant that enabled it to sponsor Mathematics Masterclasses for promising teenagers, to be mounted, in perpetuity, in the RI and countrywide.

To illustrate how major changes can occur in the RI, it is relevant to recall how this massive initiative for mathematics occurred. One day in the late 1980s, two stalwart members of the RI asked if they could come and see me, in my capacity as Director. They were Professor John Waterlow and his long-time friend—they were at Eton together as schoolboys—Colonel John Innes, a scion of the famous John Innes family. They began by telling me how much they appreciated being members of the RI, and greatly complimented me for doing what they felt was an effective job as its Director and in leading the RI appeal. But, they said, they were concerned about how inadequate British children were, mathematically, compared with their counterparts in the Far East, notably Singapore, China, and Japan. They had noted how encouraging my predecessor George Porter had been in inaugurating the Saturday morning Mathematics Masterclasses mounted at the RI in the autumn and winter months.

'*Why cannot such masterclasses in mathematics, under the aegis of the RI, be mounted in towns all over the country?*' (I remember vividly that they named Swindon and, possibly to appeal to my emotions, Blaenau Ffestiniog.) They proceeded to encourage me to do something positive to advance the skills of British teenagers in mathematics, and specifically drew my attention to the fact that they were both members of the Worshipful Company of Clothworkers. They further pointed out that if I were to submit a really ambitious proposal whereby Mathematics Masterclasses could be instigated all over the UK, their Company would look upon it seriously, provided I costed the scheme appropriately. After seeking the advice of HRH The Duke of Kent, who was President of the RI, and with the help of my colleague (and Deputy Director) the Professor of Natural Philosophy, Richard Catlow, a small team helped me draw up a submission, which led to my being fiercely interrogated for almost a whole day at the RI by four 'investigatory' Clothworkers.

Some three months or so later, I was jubilant, as were my colleagues, especially my personal assistant, Judith Wright (who had greatly helped George Porter initiate the very first round of Clotherworker-sponsored Masterclasses at the RI). Our application had been approved.

There was great excitement one summer's day in 1990 when HRH The Duke of Kent, fresh from his day at the All England Club, Wimbledon (of which he was also President), came to the RI to unveil a plaque in the Old Library on the ground floor to denote the establishment, in perpetuity of the RI Mathematics Masterclasses.

1.7   The Central Role of Mathematics in a Cosmic Context

Mathematics, more than any other scientific discipline, holds the key to understanding the large-scale properties of the physical world. This in itself is very strange, for mathematics is the free creation of the human mind bound only by the laws of logic and governed by the fertility of the human imagination. To quote Professor John Polkinghorne, who gave two Discourses at the RI, his most recent in the 1990s: '*Our mathematical friends sit in their studies and think their austere thoughts, yet the abstract structures they create can prove to fit exactly the concrete structures of the world around us.'* In the words of the current Astronomer Royal, Martin Rees, it is interesting that most of the mathematics that physicists have needed was already 'on the shelf'. To take a specific example, in 1984 the New Zealander Vaughan Jones was working in a somewhat arcane area that centres on the pure mathematics of the knotedness of knots: how to describe and interpret the properties and distinguishability of knots. During the course of that work he arrived somewhat unexpectedly at what has since become known as the Jones polynomial or the Jones invariant. His fresh insight was quite extraordinary, for it was soon to revolutionize many seemingly different branches of physics and mathematics and latterly biology. It has since transpired that the Jones polynomial is the pivot around which some of the advanced branches of modern-day physics turn. Thus, topology, which is concerned with the connectedness of visible objects (like knots in a string), was shown by Jones to be linked to statistical mechanics and specialized branches of algebra in a remarkably unexpected way. Moreover, many major areas of cosmology and modern physics, including relativity, quantum field theory, and electromagnetism have since been shown to be interrelated via the Jones polynomial. To cap it all, Jones' work in knot theory, pursued for its own sake with no thought of any practical application, has already been of significant use in molecular biology in that it offers explanations for hitherto puzzling features in the folding of the double helices of DNA, the most important molecule in all living things.

Professor John Spence,[19] another RI Discourse speaker, has written cogently on this topic, and in so doing has quoted the Hungarian-American Nobel Laureate Eugene Wigner: *'Wigner points out that, with a different kind of imagination and curiosity, mathematicians are primarily interested in problems which advance the core agenda in their subject, mathematics, not physics. They therefore tackle problems of the greatest punchy mathematical interest, unrelated to physical phenomena.'* He goes on to quote Wigner (who was echoing a similar sentiment previously expressed by Lord Kelvin, an ardent member of the RI)[20]:

'The miracle of the appropriateness of the language of mathematics for the formulation of the laws of physics is a wonderful gift which we neither understand nor deserve. We should be grateful for it and hope it will remain in future research and that it will extend for better or worse, to our pleasure, even though perhaps also to our bafflement, to wide branches of learning'.

1.8   The RI as Mecca

So highly do scientists in general, and chemists and physicists in particular, revere the individuals, especially Faraday and Davy, who worked at the RI, that an unending stream of eminent individuals regularly visit it, or opt to lecture in its historical theatre, whenever the opportunity arises. In my day as Director, dozens of individuals from all over the world made requests during their visit to the UK to be shown around the RI, just to savour the historicity and also the sense of the genius of the place where Faraday and Davy worked and lectured. This desire to visit the Mecca of physical science is apparent among visiting scientists the world over. But among Japanese savants the desire seemed to be greatest of all. Photographs of some of the distinguished Japanese scientists, which I alone showed *around the precincts of the RI, are shown* in Figure 1.16.

Professor Kenichi Fukui, ForMemRS, and Nobel Prizewinner in Chemistry (shared with Professor Roald Hoffmann, who gave a Friday Evening Discourse in the RI in 1987), was especially proud to attend a Discourse at the RI in 1991. His fellow countryman Professor Kenzi Tamaru, former Head of Chemistry at the University of Tokyo, and a pioneer catalytic chemist of Tokyo, relished his visit to the RI, and in particular its lecture theatre in 1988, and so did his successor as Head of Chemistry in Tokyo, Professor Hario Kuroda, the prime driving force behind the establishment of the Spring-8 synchrotron in Japan (one of the foremost in the world), and the distinguished electrical engineer Professor

(b) Professor Kenzi Tamaru, Head of
Chemistry, University of Tokyo

(a) The Nobel Laureate,
Kenichi Fukui, ForMemRS

(c) Professor Hiroshi Inose, eminent
electrical engineer and founder of the
Institute of Informatics, Japan

(d) Professor Hario Kuroda, principal architect of one of the most
advanced synchrotron sources in the world, the SPring-8

Figure 1.16 *Photographs of eminent Japanese scientists who came to visit the RI in my days as Director (1986–91). Top centre, copyright Professor Kenzi Tamaru; the other three, courtesy Mrs Fukui, Mrs Kuroda, and Mrs Inose.*

(Personal permissions JMT)

Hiroshi Inose, who invented the time-slot and later established the National Institute of Informatics, Japan, and who was extraordinarily impressed by the Discourse that he attended in 1992 entitled *'Unpredictability and Chance in the Evolution of Science and Technology'*.

One of the most lucid of present-day writers (not only on scientific topics), Professor Richard Dawkins has captured perfectly, in his autobiography *'Brief Candle in the Dark: My Life in Science'*[18], the feelings that overtake intending lecturers and visitors when they enter the RI:

> *'...The Director of the RI in London was ringing to invite me to give the Royal Institution Christmas Lectures for Children, and I went hot and cold as he did so. The warm flush of pleasure at the honour was swiftly followed by a cold wave of trepidation. I immediately knew I would not be able to refuse the commission, and yet I lacked confidence that I could do it justice. I was aware that the renowned series of lectures had been founded by Michael Faraday...'*

When academician Kirill Zamaraev, Director of the Boreskov Institute of Catalysis in Novosibirsk (under the auspices of the Russian Academy of Sciences), was offered venues for his Centenary Prize Lectureship of the Royal Society of Chemistry in the early 1990s, he opted to give his lecture at the RI, from where Faraday spoke on more than a thousand occasions.

When Faraday passed away, the Chemical Society (as it was before it became the Royal Society of Chemistry, a century or so later) in 1867 instituted the Faraday Prize and Lectureship, still its premier award. Every three years, one of the most famous physical scientists in the world is awarded this prize, which requires the recipient to give a lecture in Britain, at a venue which they are free to choose. Table 1.4 gives the list of recipients of the prestigious Faraday Prize for the first hundred years since it was inaugurated. The vast majority of these recipients chose to give their prize lecture at the RI.

Eighteen of these Prizewinners—there have been more subsequently—were Nobel Laureates. Up until 1900 there were no Nobel Prizes, yet each Faraday Prizewinner from 1869 until the end of the century were among the most famous scientists in the world. And they each commenced their lectures at the RI, at Faraday's lecture desk, lavishing praise on the eponymous hero.

A selection of the nature of topics and quality of the speakers at Discourses in the RI from the early 1930s to the mid-1960s is enumerated in Table 1.5.

Visitors whose interests and work have been largely in the physical sciences and in general education are enormously influenced by the mark that Faraday left on the RI. We return to this topic again in Chapter 4, but it is relevant to emphasize that, as acknowledged by a succession of indisputable authorities, Faraday was arguably one of the greatest experimentalists and natural philosophers ever—comparable to Mozart or Haydn in music. He bequeathed to posterity a larger and more widely varied corpus of scientific information than any other physicist or chemist. In Chapter 4, we shall see why Albert Einstein believed that Faraday,

Table 1.4 *Recipients of the Faraday Prize and Lectureship for the first 100 years of its existence (eighteen of these were Nobel Prizewinners).*

1869	Jean-Baptiste Dumas	French
1872	Stansiloo Cannizzaro	Italian
1875	August-Wilhelm von Hoffmann	German
1879	Charles-Adolphe Wurtz	German
1881	Hermann von Helinholtz	German
1889	Dmitri Ivanovich Mendeleev	Russian
1895	John Strutt, Third Baron Rayleigh	British
1904	Wilhelm Ostwald	German
1907	Hermann Emil Fischer	German
1911	Theodore William Richards	American
1914	Svante Arrhenus	Swedish
1924	Robert A. Millkan	American
1927	Richard Willstätter	German
1930	Niels Bohr	Danish
1933	Peter Debye	Dutch
1936	Ernest Rutherford	New Zealander
1939	Irving Langmuir	American
1947	Sir Robert Robinson	British
1950	George de Hevesy	Hungarian
1953	Sir Cyril Hinschelwood	British
1956	Otto Hahn	German
1958	Leopold Rijieka	Yugoslavian
1961	Sir Christopher Ingold	British
1965	Ronald G. W. Norrish	British
1968	Charles Coulson	British
1970	Gerhard Herzberg	Canadian

along with Maxwell, gave rise to a total revolution in humankind's concept of the physical world.

There is one final incident that I am prompted to cite to illustrate how much of a Mecca the RI continues to be. When the book entitled *'Michael Faraday and the Royal Institution: The Genius of Man and Place'*[8] was published by the Institute of Physics in the summer of 1991, that organization (and myself, as author) was immediately approached by a distinguished scientist from the *Japan Association for International Chemical Information* in Tokyo. This scientist, Professor Hideo Chihara, said that his association had commissioned him to translate my book into

Table 1.5 *A Selection of some of the Discourse speakers and their topics in the 1930s, 1950s, and early 1960s.*

Date of Discourse	Speaker	Topic
01/05/1931	D'Arcy Thompson	Charlotte Brontë in Brussels
17/02/1933	J. Dover Wilson	The plot of Hamlet—a rediscovery
26/05/1933	Sir Walford Davies	Pure music and applied
23/02/1934	Charles Morgan	A defence of storytelling
20/04/1934	P. M. S. Blackett	Cosmic rays
01/03/1935	Sir Arthur Bryant	Samuel Pepys
11/02/1938	Archbishop William Temple	Truth in science, poetry, and religion
31/03/1939	Mary Somerville	The broadcasts for schools
26/11/1940	Kenneth Clark	The cinema as a means of propaganda
03/12/1940	J. D. Bernal	The physics of air raids
05/03/1942	E. M. Forster	Virginia Woolf
13/11/1942	Lord David Cecil	Turgenev
15/05/1953	Edwin Hubble	The observational evidence for an expanding universe
17/05/1957	Kathleen Kenyon	Excavations of Jericho
29/04/1966	Fred Hoyle	Quasi-stellar objects
20/05/1966	Murray Gell-Mann	Elementary particles

Japanese, which naturally pleased me, and it was duly published a year later.[21] During the course of Professor Chihara's work of translation, we exchanged numerous letters. But, what was significant, in the context of the RI's power of attraction, was that he once came to visit it in London because, he said, *'it occupied a unique place in the history of science.'*

REFERENCES

1. Kathleen Coburn, 'Coleridge, a Bridge Between Science and Poetry: Reflections on the Biochemistry of his Birth', *Proc. Roy. Inst. Great Brit.*, **1973**, *46*, 45.
2. Coleridge was among the first—some claim he was the first—to use the word 'psychosomatic', which did not become a serious subject of medical study until a few decades later.
3. Oliver Sacks, the British-American neuroscientist and author, has written a fascinating chapter on Mendeleev in his book *'Uncle Tungsten: Memories of a Chemical Boyhood'*, Picador, London, **2001**.

4. Young built an apparatus consisting of a water tank in which there were two sources of vibration as a means of showing the reality of the phenomenon of interference of wave motion, as an analogy for the interference of light waves produced in his 'double-slit' experiment.

5. I am informed by expert Egyptologists that Champollion never disclosed (or confessed?) to reading Young's definitive article on Egypt published in 1818. There is no doubt, however, that Champollion went much further than Young in deciphering hieroglyphics. (See Chapter 9.)

6. Both the Clarendon Laboratory of physics at the University of Oxford and its counterpart the Cavendish Laboratory at the University of Cambridge were established in the early 1870s and soon became two of the premier research laboratories in the world.

7. Thomas Martin, '*The Royal Institution*', Royal Institution, London, **1961**.

8. John Meurig Thomas, '*Michael Faraday and the Royal Institution: The Genius of Man and Place*', IOP Publishing, Bristol, **1991**.

9. 'Michael Faraday', *Illustrated London News*, **1861**, *28*, 38.

10. A list reproduced from A. F. Jeffrey, '*Michael Faraday. A Fest of his Lectures and Published Writings*', Royal Institution, London, **1960**.

11. See p. 38 of ref [8] above.

12. Sir William Crookes, inventor of the radiometer and of the evacuated tubes named after him, and the discoverer of the element thallium. At various times he was president of the Chemical Society, the Society of the Chemical Industry, the Institute of Electrical Engineers, the British Association, and the Royal Society.

13. Z. Su, W. Zhou and Y. Zhy, *Chemical Communications*, **2011**, *47*, 4700.

14. S. P. Thompson, '*Light, Visible and Invisible*', Macmillan, **1897**.

15. W. H. Bragg, '*Concerning the Nature of Things*', Bell, **1924**.

16. W. H. Bragg, '*Old Trades and New Knowledge*', Bell, **1926**.

17. R. Dawkins, '*River Out of Eden: A Darwinian View of Life*', Basic Books, New York, **1995**.

18. R. Dawkins, '*Brief Candle in the Dark: My Life in Science*', Bantam Books, New York, **2015**.

19. J. C. H. Spence, '*Lightspeed*', Oxford University Press, Oxford, **2009**.

20. E. Wigner, 'The Unreasonable Effectiveness of Mathematics in the Natural Sciences', in '*Communications in Pure and Applied Mathematics*', J. Wiley & Sons, New York, **1960**, p. 13.

21. It has subsequently been translated into Italian and Chinese.

2

Count Rumford and his Remarkable Creation in Albemarle Street

2.1 Introduction

On 19 September 1798, there arrived in London from Munich a forty-five-year-old, tall, blue-eyed, handsome, American-born opportunistic man of action. A former soldier and statesman, he was, by all accounts, ruthless and arrogant, callously cunning and devious, an accomplished spy, and a calculating womaniser. But he was also a philanthropist, a brilliantly effective social reformer, an ingenious inventor, and an exceptionally innovative scientist. His name was Benjamin Thompson, better known as Count Rumford of the Holy Roman Empire. He was the man who established the Royal Institution (RI).

Born in 1753 in Woburn, Massachusetts, to simple farmer folk, Rumford Thompson, founded many things, including the Rumford Professorship at Harvard University, and the Rumford prizes and medals of both the Royal Society, London, and the American Academy of Arts and Sciences. The RI owes its origins entirely to Rumford: his career and character determined its original form. And the story of how it came to be created, what functions it was originally intended to serve—touched upon in Chapter 1—as well as the interaction between its founder and Sir Joseph Banks, President of the Royal Society from 1778–1820, is a fascinating one.

It was the brainchild of Rumford, 'whose obscurity' nowadays in North America 'is widespread'. Yet, according to US President Franklin D. Roosevelt, Rumford, along with Benjamin Franklin and Thomas Jefferson, *were the three greatest intellects America ever brought forth*.

Rumford was something of a scoundrel and undoubtedly an incorrigible opportunist. When contemplating the many and varied actions of Rumford, one recalls Bertrand Russell's remark that *'man is a strange amalgam of saint and sinner'*. In Rumford's case there was more sinner than saint. He relented and mellowed somewhat in later life, but throughout his career he devoted considerable energy to evolving ways of bettering the conditions of the poor and adding to the comforts

Albemarle Street: Portraits, Personalities, and Presentations at the Royal Institution. John Meurig Thomas,
Oxford University Press. © Sir John Meurig Thomas 2021.
DOI: 10.1093/oso/9780192898005.003.0002

of mankind, although to quote Cuvier,[1] *'it was without loving or esteeming his fellow creatures that he had done all these services.'* He was, however, an outstanding natural philosopher who uncovered many universal truths. It is worth recalling that truth is independent of the stimulus that has provoked its discovery, and the conditions that have guided its expression.

In an age when the natural philosopher tended to belittle the engineer and treat him as an artisan or common labourer, and the inventor, in turn, regarded the natural philosopher as an intellectually snobbish dreamer, Rumford wrote, argued, and practised—well ahead of his contemporaries—a scientific philosophy which cogently stated that fundamental research in a problem is a prerequisite to technological development. Furthermore, Rumford described his work with elegance and charm: he was a beautiful stylist who introduced a perspective and an aura of Chekovian timelessness to his discourses. A good example is the opening of his important paper, which has become a classic in the annals of science, 'An Inquiry Concerning the Sources of the heat which is excited by friction' (*Phil. Trans. R. Soc. Lond.*, 1798, 80): *'It frequently happens, that in the ordinary affairs and occupations of life, opportunities present themselves of contemplating some of the most curious operations of nature; and very interesting philosophical experiments might often be made, almost without trouble or expense, by means of machinery contrived for the mere mechanical purposes of the arts and manufacturers.'* Rumford continues in this vein and expatiates on *'the playful excursions of the imagination'*, which led him inexorably to the conclusion, described in this paper, that the boring of cannon generates an inexhaustible supply of heat. Rumford also knew that scientific research is a passionate undertaking and he expressed his feelings in terms that ring true with all those who have been gripped by obsessional preoccupation: *'The ardour of my mind is so ungovernable that every object that interests me engages my whole attention and is pursued with a degree of indefatigable zeal which approaches to madness.'* (My late friend, Max Perutz, a great RI man, relished this statement by Rumford.)

2.2   Early Life: Soldier, Statesman, and Scientist

In rather quick succession in his middle teens, Benjamin Thompson tried his hand at commerce, medicine, and teaching; he also attended some courses at Harvard, and in July 1772 he took up a position as schoolmaster in the town of Concord (formerly called Rumford), New Hampshire. Four months after arriving there, he married the wealthiest widow in the province. She was fourteen years his senior. He thereby became not only a wealthy gentleman-farmer but gained entry into the political and social circle surrounding the Royalist Governor, John Wentworth.

He soon became extremely unpopular in Concord because of his overt espousal of the American Tory cause in the gathering strain between the British Crown and the American Colonies. So, less than eighteen months after his arrival in Concord, he fled to Boston to serve General Gage as a Tory spy.[2] His wife, with their baby daughter in her arms, was left to the mercy of the angry citizens. He never saw his wife again. In 1776, after having been arrested for spying, then released owing to a lack of evidence, he sailed for London to report directly to King George III on the situation in America.

This self-assured, twenty-three-year-old expert on the American Army around Boston, holding the rank of major in the New Hampshire Militia, was made private secretary to the Secretary for the Colonies. He ingratiated himself with the aristocracy, and he was soon dining and relaxing in the stately homes of England.[3] But he also began experimenting in the field of natural philosophy at Stoneland Lodge, Sussex, a country seat of Lord George Germain. It was in the grounds of Stoneland Lodge, assisted by the Reverend Bale, Rector of Withyham, that Thompson undertook his first real experimental investigation, published in *Philosophical Transactions*.[4]

He developed a technique based primarily on the ballistic pendulum method devised by Benjamin Robins some forty years earlier. Robins' experiment was to fire a missile from a small cannon into a heavy wooden block supported as a pendulum and to measure the momentum exchange between the bullet and the heavy pendulum. Using the principle of the conservation of momentum, the velocity of the bullet as a measure of the force of the gunpowder could be computed. Thompson's improvement was to argue that the windage around the ill-fitting bullets of the day caused large errors in a measurement taken in this manner and that measuring the recoil velocity of the gun was a great improvement.

This experimental study so impressed his scientific colleagues, especially the influential President of the Royal Society, Sir Joseph Banks, that Thompson was elected Fellow of the Royal Society (FRS) in March 1779.

In 1780, he was made Under-Secretary for the Northern Department in the British Government, a post which put him directly in touch with the American scene. The following year, he abruptly left his post and sailed for America as Lieutenant Colonel of the King's American Dragoons. His conduct and general behaviour in his native land was nothing short of despicable.

He returned to England at the close of the Revolution, transferred his regiment to the Regular British Army, and applied for, and obtained, the rank of full colonel. This secured for him a pension for life at half pay. In 1783, he sat for his portrait, which now hangs in the Fogg Art Museum, Harvard University, and became the only American-born to be painted by Thomas Gainsborough.

2.3 Bavarian Adventures

Almost immediately thereafter, he left[5] London to seek his fortune in Europe, and he entered the service of Karl Theodor, the Elector of Bavaria, and settled down to his most productive and spectacularly successful years of invention, scientific endeavour, and social reconstruction. While negotiating with Karl Theodor to

Figure 2.1 *Thomas Gainsborough's portrait of Sir Benjamin Thompson, later Count Rumford (1753–1814).*

(Reproduced by kind permission of Harvard Art Museums/Fogg Museum, bequest of Edmund C. Converse. Accession number: 1922.1; photograph credit: President and Fellows of Harvard College)

become his aide-de-camp, he realized that his rank in the Bavarian Court would exceed his standing as a British subject. This presented him with another opportunity for advancement. In a quick visit to London he persuaded King George, for whom he had earlier agreed to spy on the Bavarians, to confer upon him a knighthood, so that it was as Colonel Sir Benjamin Thompson that he entered the foreign prince's service.

His actual job in 1784, at the outset of his eleven-year stint in Munich, was to advise the Elector of ways of reorganizing and modernizing the Bavarian Army. He negotiated for himself a highly favourable financial deal if he succeeded to increase the efficiency, morale, and fighting ability of the army. He quickly identified that the major item of expense in the military budget was for clothing. As a result, he turned his attention towards the basic physics of insulation, so that he could subsequently direct the manufacture of more efficient protection for the soldiers. To this end, he invented the so-called cylindrical passage thermometer, which he used to discover that the insulating properties of a cloth results from the air caught in its interstices. (In the process of using his cylindrical passage thermometer, Thompson also discovered convection currents.)

Having solved the fundamental physics of the problem, he set about trying to manufacture cloth which conformed to his discovered principles, but no manufacturer in Munich and its environs was interested in such an advanced idea. He decided to set up his own factory to make clothes for the Bavarian Army, but the labour force was simply not there. So he turned to a radical solution which, at the same time, transformed the social fabric of Munich. In those days, the city was overrun with beggars and vagrants—one in thirty of all adults were mendicants. On 1 January 1790, using the dictatorial powers invested in him by the Elector, he ordered the city garrison into the streets to arrest every one of the tatterdemalion horde in Munich and threw them into the great stone city prison, which he promptly renamed a House of Industry. All the beggars and their wives and children were employed in making uniforms for the Bavarian Army. He clothed them, housed them comfortably, and fed them well, using his own recipe for soup, a premier constituent of which was the potato,[6] which he was responsible for introducing as a staple food. (Other products of Thompson's fertile mind that emerged during his period in Munich as the Elector's supremo are listed in Table 2.1)

He was fascinated by the whole technology of cooking. Indeed, many regard him as a founding father of domestic and culinary science. He was, for example, the first to introduce the use of baking soda; he was also responsible for the first really efficient fireplaces and chimneys.

In the House of Industry in Munich, the small amount of natural light that was available resulted in great inefficiency among the workers, who worked up to fourteen hours a day.

Thompson therefore turned his attention to increasing the efficiency of lamps and candles. First, he had to devise a method of measuring the intensity of light,

Table 2.1 *Rumford as inventor and technologist*

- Kitchen range
- Pressure cooker
- Convection oven
- Rumford stove
- Rumford lustre pots
- Drip coffee maker
- Fireplace damper
- Double-pane glass
- Efficient oil lamps
- Photometer
- Cylindrical passage thermometer

and the photometer which he invented still bears his name. The candle used as a standard in his experiments on luminosity he defined very carefully. Indeed, his definition of the standard candle was used[2] for more than a hundred years as the international standard candle. He so improved the oil lamps of the day that many examples of his improved lamps are still to be found in antique shops and museums.[7]

Having discovered convection currents, Thompson properly analysed the requirements of the control of hot and cold gases in fireplaces and chimneys. Prior to his innovations, chimneys were perfectly straight holes from the fireplace upwards. By introducing a smoke shelf and a throat to the chimney, cold air comes down the back of the chimney and up its front without turbulence.

Another of Thompson's fundamental experiments was the study of radiation from black and shiny surfaces. He demonstrated that good reflectors were poor radiators and vice versa. (While on the subject of his practical inventions, we should also note his contributions to pure science, some of which were to come later in his career. These are listed in Table 2.2).

By housing, feeding, and clothing the poor of Munich, he incorporated their endeavours constructively into the scheme of things and enhanced social stability and contentment. His approach to social reform is captured in the following passage written by him[3]:

'To make vicious and abandoned people happy, it has generally been supposed necessary first to make them virtuous. But why not reverse the order? Why not make them first happy, and then virtuous?'

In this spirit, the pleasure grounds and park with artificial lake, refreshment saloons, and Chinese pagoda known as the English Garden in Munich, was designed and created by Thompson in 1790. It became and remains attractive as a retreat and sanctuary to all strata of Bavarian society.

Table 2.2 *A selection of Rumford's contributions as a scientist*

- Demonstration of the nature of heat
- Measurement of heats of combustion of fuels (and devising a calorimeter to do so)
- Development of techniques for measuring intensity of radiation and defining 'candle power'
- Determination of properties of gunpowder and contributions to ballistics
- Measurement of specific heats of many materials
- Discovery of convection currents in liquids
- Measurement of thermal conductivity of many gases and heat lost by radiation from surfaces

In 1791, Sir Benjamin was so powerful that he was simultaneously Minister of War, Minister of Police, Major-General, Chamberlain of the Court, and State Counsellor. In the following year, during an interregnum in the succession of emperors of the Holy Roman Empire, Elector Karl Theodor was Vice-Regent, and, having for a short while the power to create nobility himself, conferred on his brilliant minister the title of Count Rumford.

Rumford's amorous adventures in Munich, as elsewhere, were complicated and numerous. Countess Baumgarten, the mistress of the Elector, became a long-standing companion and she bore Rumford a daughter. Her sister Countess Nogarola also became his mistress. Rumford visited her frequently in Verona and other parts of northern Italy, where he also had an affair with the English aristocrat Lady Palmeston, whom he first met in Milan. For five years in Munich he had a young housekeeper and mistress, Victoire Lafevre, who later bore him a son when he was sixty years of age.[8]

Throughout his time in Bavaria, Rumford's irrepressible scientific curiosity was fully exercised, especially in regard to the nature of heat. Through his observations of the drillings of cannons in the arsenal that he had built, he concluded that heat is not an igneous fluid (as he put it) which flows in and out of bodies, but was an immaterial substance. The drilling of cannon could clearly deliver an inexhaustible supply of heat. He thereby demolished the caloric theory of heat, although it still had its champions for a decade or so after Rumford's paper appeared in *Philosophical Transactions*.

In 1795, he took a two-year leave of absence from the Elector to enable him to return to England to publish his essays[9] and to communicate his important results to the Royal Society, Fellowship of which he greatly valued. These two years in London were the happiest of his life. He had short trips to places such as Dublin and Edinburgh, where he was festooned with honours, as well as Bath and Harrogate, living the life of a famous natural philosopher and philanthropist.

During his stay in Harrogate, Rumford made careful experiments on himself with regard to the warm bath. These are given in his thirteenth essay, in the 'Salubrity of Warm Bathing' (see Table 2.3).

He found that a daily bath of ninety-six or ninety-seven degrees for half-an-hour, two hours before dinner, for five weeks, increased the appetite, the digestion, the spirits, the strength, and the insensibility of cold. To illustrate the diversity of interests that Rumford exhibited in making people more comfortable and societal life more effective and cleaner, we show in Table 2.3 most of the contents of Volume III of *'The Collected Works of Count Rumford'*. [10]

An important practical achievement which earned him enhanced admiration at that time was the improvement he made to domestic heating. As stated earlier, his studies of convection currents and heat radiation led him to revolutionize the design of fire-places and chimneys, and he claimed to have modified more than two-hundred-and-fifty fireplaces in London alone in a single two-month period in 1796. His fame led to a famous James Gillray cartoon, showing Rumford in a state of almost maniacal contentedness, standing in front of his stove. (In Jane Austen's novel *'Northanger Abbey'* (1817), General Tilney is described standing in front of a Rumford stove.)

Table 2.3 *Examples of the diversity of interest of Count Rumford (taken from the contents of* 'The Collected Works of Count Rumford'[10])

1. 'Use of Steam as a Vehicle Transporting Heat'
2. 'Description of a New Boiler Constructed with a View to the Saving of Fuel'
3. 'On the Management of Fires in Closed Fire-Places'
4. 'On the Construction of Kitchen Fire-Places and Kitchen Utensils'
5. 'On the Salubrity of Warm Bathing'
6. 'On the Specific Gravity, Strength, Diameter, and Cohesion of Silk'
7. 'An Account of Some Experiments Made to Determine the Quantities of Moisture' Absorbed from the Atmosphere by Various Substances'

In 1796, he made his famous and generous bequests to the Royal Society of London and the American Academy of Sciences in Boston. He gave to each body $5,000 to finance the award of a medal for outstanding work done in the fields of heat and light.

However, with Rumford's dynamic personality out of the way in England, his numerous adversaries in Munich worked unceasingly to undermine his power so that when he returned to Bavaria it was obvious that his days were numbered. Even the Elector decided to forsake him, but he did so by letting him go with a great honour: as Minister Plenipotentiary to the Court of St James, in England. Karl Theodor did this without first consulting George III, who, it is alleged, was incandescent with rage when Rumford appeared to present his credentials. King George refused to have as a foreign minister one of his own beknighted subjects who, it had transpired earlier, had been accused of spying (for France) on his government. And so here was Rumford in 1798, aged forty-five and temporarily unemployed in London.

2.4   West Point or Albemarle Street?

Rumford set many schemes in motion, one of them being to offer his services to US President John Adams as the Superintendent of the Military Academy at West Point. At first, Rumford's suggestion was accepted by Adams. But months later, after some delicate investigations by Rufus King, the American Ambassador, it was discovered that he had spied on the American Army. Great manoeuvring then went on so that all parties could save face: Count Rumford was formally invited to come and accept the position after proper assurance had been given that he would decline the honour.

Yet, Rumford still had many influential friends in London, among them the powerful President of the Royal Society, Sir Joseph Banks; Henry Cavendish; the second Earl Spencer (one of Diana, Princess of Wales's forebears); William Wilberforce, the social reformer; and Thomas Bernard, a rich former barrister,

Figure 2.2 *The James Gillray cartoon entitled 'Comforts of a Rumford Stove' showing Count Rumford standing in front of one of his fireplaces.*
(Copyright Capra)

whose entire life centred around philanthropy. It was thus decided at a meeting in Banks's home in Soho in March 1799 that Rumford should draw up the plans for forming a new institution, the RI. And so we have his proposals:

> *'...for forming by subscription, in the Metropolis of the British Empire, a Public Institution for diffusing the knowledge and facilitating the general introduction of useful mechanical inventions and improvements, and for teaching by courses of philosophical lectures and experiments the application of sciences to the common purposes of life.'*

Note, in particular, the aims to:

- spread a knowledge of new and useful mechanical improvements
- teach the application of scientific discoveries to the improvement of arts and manufactures and to the increase of domestic comfort and convenience

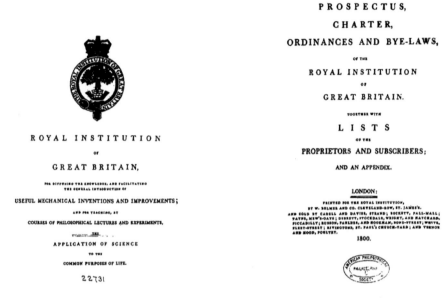

Figure 2.3 *A copy of the prospectus distributed by Rumford and his colleagues shortly after the Royal Institution (RI) was founded.*

(Courtesy American Philosophical Society)

Note, too, there was no mention whatsoever of pure science in Rumford's vision. In his RI, science was primarily intended to make itself useful. His first aim was to be achieved by setting up a permanent exhibition consisting of what one would now regard as a hybrid between a museum of science and a world fair in miniature. The second was to be achieved by lectures on applied science backed by a laboratory for chemical and other experiments.

Rumford's proposals, of which many hundreds of copies were published, were given wide publicity and subscriptions were invited. All who gave fifty guineas or more were to become perpetual Proprietors of the RI. Fifty-eight influential people quickly agreed to subscribe, including one duke, six earls, seven lords, eleven knights, one bishop, and eighteen Members of Parliament.

Rumford devoted all his considerable energies and acumen to his new foundation and worked indefatigably on its behalf. He framed its constitution and saw to it that a Royal Charter was solicited, and that a house in Albemarle Street, in the heart of London's West End, was purchased and laid out according to his plans, with a capacious lecture theatre, a repository (or model room) for the exhibition of mechanical inventions, including stoves and fireplaces, and with kitchens and workshops. In due course, several additions were made, including a library and reading room.

As its first Professor, Rumford enticed the estimable physician and scientist Thomas Garnett from Anderson's Institution, which had been set up in Glasgow two years earlier. When the dynamic Rumford was present at the RI in Albemarle Street the whole place hummed with a feverish activity. But he worked so hard that he was absent ill for five months. In the time that he was away, dissent festered and Sir John Coxe Hippisley and his fellow managers began to ventilate their fundamental differences of opinion with Rumford. In early 1800, Rumford returned with renewed vigour, assumed total control of the RI, and lived in the house, where he held weekly progress meetings and galvanized all and sundry into action.

But some of the other officers thought Rumford dictatorial and overbearing; they resented his bullying and felt that his whole manner was irksome. Rumford's financial profligacy troubled Bernard profoundly, and soon these two protagonists were at daggers drawn. Moreover, Rumford's plans and convictions increasingly fell foul of the other managers. And prominent industrialists such as Matthew Boulton[11] felt it would be ruinous to enterprising manufacturers and inimical to British interests if every spectator were to be allowed to examine useful equipment in the proposed repository.

To compound Rumford's difficulties, Karl Theodor died suddenly in 1799, so he felt it necessary to return briefly to Munich to be reassured about his pension by Maximillian Joseph, the Elector's nephew and successor. On his return journey, he went to Paris for a vacation and was so well received (by Napoleon, Pierre-Simon Laplace, Joseph-Louis Lagrange, Talleyrand, and others, and especially by Madame Lavoisier) that he stayed there for two months. His prolonged absences from the RI and the increasing ascendancy of his adversaries there distressed Sir Joseph Banks, who told Rumford that the RI was in the hands of the profane.[12]

Meanwhile, Rumford had fallen out with Thomas Garnett, whom he treated with a singular lack of understanding and no scintilla of compassion. Garnett soon resigned. But before that happened, Rumford had identified a twenty-three-year-old Cornishman then working in the Pneumatic Institute, Bristol, and who was destined to become the brightest star in the European scientific firmament for the next twenty years—Humphry Davy (as stated in Chapter 1). Davy was appointed a lecturer; and in Garnett's place as Professor, Rumford chose another West Country Englishman, Thomas Young, of Young's modulus fame. It is interesting to record that Davy had soon tried to convince the powers that be at the RI (see Chapter 3) to consider mounting courses on medicine.[13] It must not be forgotten that several of the early members of the RI—Edward Jenner of the smallpox vaccine fame being one of them—were medical men of leading rank.[14]

But things were proceeding so much at variance with his proposals at the RI that Rumford openly declared, after a three-month visit to Munich and Paris from October 1801 to January 1802, that he would leave it and England for France, forever.

2.5 Madame Lavoisier

The attraction in Paris was a famous wealthy widow, Madame Lavoisier, whose distinguished husband had been guillotined in the French Revolution. Rumford and Lavoisier saw in each other a chance to settle down to an idyllic life of amusement and pleasure. After encountering many initial legal difficulties, eventually they married in October 1805. But there were immediate marital difficulties,[15,16] which led to permanent separation in 1807. Acrimonious and vituperative rows between them flared up in public to the embarrassment of their hosts or guests. She went so far as to sabotage his garden. He fled from the centre of Paris to his new abode on its outskirts, where, in his house in Auteuil, he carried out scientific work of far-reaching importance, until the time of his death there in 1814.

In 1806–7, Rumford sought relief through the pursuit of science. Although his personal relationships were miserable, he published eight papers in the *Memoirs of the Institute*:

- a description of a new differential thermometer
- research on heat, showing the effect of difference of surface on radiation
- further experiments on the effect of blackening the surface
- different properties of bodies with respect to radiation and to conducting powers
- the passage of heat through solids
- on the heat of solar rays
- on the cooling of liquids in vases of porcelain, gilt or not gilt
- dispersion of the light of lamps by serceus of ground glass, silk, and so forth with a description of a new lamp

His experiments on the temperature of water at the maximum density, which he also carried out during this period, he published in *Phil. Trans. R. Soc. Lond.*

At the termination of Rumford's life, what could be said of his remarkable creation, the RI? Paradoxically, pure research, the very activity that Rumford had excluded from his proposals, was to prove one of its principal saving graces. The other was the personalities and genius of Davy and Michael Faraday, whose lives, work, and legacies we discuss in Chapter 3 and Chapter 4.

REFERENCES

1. Baron Georges Léopold Chrétien Frédéric Dagobert Cuvier (1769–1832), French anatomist and taxonomist.
2. There is proof of the excellence of Thompson's scientific techniques in the secret ink which he used to communicate to the blockaded British Army in Boston. He used

gallotannic ink developed by ferrous sulphate, which is said to have been as good as many of the secret inks used during the First World War.

3. *'The Memoir of Sir Benjamin Thompson'*, by G.E. Ellis, published in 1871 (Boston) by the American Academy of Arts and Sciences, contains an interesting description (pp. 56–57) of Thompson's oleaginous tendencies: *'Young Thompson was essentially a courtier. He manifested in early manhood the tastes, aptitudes, and cravings which prompt their possessor, however humbly born, and under whatever repression from surrounding influences, to push his way in the world by seeking the acquaintances and winning the patronage of his social superiors, who have favors and distinctions to bestow.'* This passage refers to Thompson's period in Concord and Boston (1773–5).

4. B. Thompson, 'New experiments upon gun-powder with occasional observations and practical *influences*', *Phil. Trans. R. Soc. Lond.* **1781**, *LXXI*, 230.

5. A fellow voyager was the historian Edward Gibbon (1737–94) of *'Decline and Fall of the Roman Empire'* fame, who had just lost his sinecure at the Board of Trade. In a letter to Lord Sheffield describing the crossing, Gibbon spoke of *'the soldier, the philosopher and the statesman Thompson'*.

6. Rumford soup is still talked of and prepared in continental Europe.

7. The trouble with the oil lamps of his day was that the oil was very thick and viscous, so that the lamp burned brightly at first, when filled, but was almost useless when nearly empty. Thompson overcame this problem by simply placing the oil reservoir at the same height as the wick.

8. He had many other affairs and mistresses. Princess Taxis, Baroness deKalb, and Madame Laplace, among others, were associated with him at various times.

9. Rumford's essays were published at different times between 1796–1802. His first gave an account of an establishment for the poor in Munich; the second was on establishments for the poor in general; the third was on 'Food and Feeding the Poor: Rumford Soup and Soup Kitchens'; the fourth on 'Chimney Fire-Places'; the fifth on 'Several Public Institutions Founded in Bavaria'. There were several others dealing with such topics as 'Management of Fire and the Economy of Fuel' and 'Propagation of Heat in Fluids'.

10. S. C. Brown, ed., *'The Collected Works of Count Rumford'*, Vol. III., *'Devices and Techniques'*, The Belknap Press of Harvard University Press, Cambridge, MA, **1969**.

11. Matthew Boulton, who partnered James Watt's ventures with the steam engine, was quite scathing in his criticism of Rumford's grandiose plans concerning the Repository with its exhibits.

12. What Sir Joseph Banks thought of the RI, which had mounted lectures by the poet Samuel Taylor Coleridge, by the painter Sir Edwin Landseer, and by the religious divine Reverend Sydney Smith, may be gauged from his letter to Rumford in April 1804: *'Your not appearing in England last year…has been a material disappointment to me and a great detriment to the Royal Institution.'*

13. I. M. McCabe, 'The Physicians cum Natural Philosophers at the Royal Institution, 1799–1840', *Proc. Roy. Inst. Great Brit.*, **1988**, *60*, 99.

14. The founders of the Medical Society of London (established in 1773) were all member of the RI.

15. There was one serious difficulty with Rumford and Madame Lavoisier's plans. Rumford was a British colonel and Napoleon had no wish to have such people, especially those who had a reputation for espionage activities, within the bounds of the French Empire. The couple, therefore, had to spend most of their time in Bavaria,

travelling around other countries of Europe, hoping for the First Consul to change his mind, which he ultimately did by the spring of 1804, when, to his horror, Rumford encountered another hurdle. Under French law, Rumford had to produce documentary proof that his first wife was dead. It took more than a year, owing to the hazards of war, for Rumford to be able to produce the required documents to enable them to be married.

16. Writing to his daughter from Paris on 24 October 1806, on the first anniversary of his marriage to Madame Lavoisier, he says: '*I am sorry* to say that experience only serves to confirm me in the belief that in character and natural propensities Madame de Rumford and myself are totally unlike, and never ought to have thought of marrying…Very likely she is as much disaffected towards one as I am towards her. Little it matters with me, but I call her a female Dragon—simply by that gentle name!' A year later, from his home in Rue d'Anjou, Paris, he writes in ever-more despairing terms to '*My Dear Child*', and his letter ends by saying '*After that* she goes and pours boiling water on some of my beautiful flowers.'

3

Sir Humphry Davy: Natural Philosopher, Discourser, Inventor, Poet, and Man of Action

3.1   Introduction

The gentleman whose life, work, and legacy we shall focus upon in this chapter was born in Penzance, Cornwall, in December 1778 and died in Geneva, Switzerland, shortly after his fiftieth birthday. Into his short life he packed an astonishing array of achievements. An enormous amount of literature is available pertaining to Sir Humphry Davy: two fascinating books have dealt in detail with his life and work—one by Sir Harold Hartley in 1966, and another by David Knight in 1992 entitled *'Humphry Davy: Science and Power'* (Cambridge University Press). There have been several others, including in 1839 the comprehensive *'The Life of Sir Humphry Davy, Bart.,'* by his brother John Davy (Smith, Elder and Co., Cornhill); the 1980 collected works edited by Sophie Forgan, *'Science and the Sons of Genius: Studies on Humphry Davy'*; and a compilation and discussion in 1990 of Davy's published works by J. Z Falmer (American Philosophical Society). One of the most illuminating accounts of the significance of Davy, in science generally, is that which has recently appeared in the posthumous publication of Oliver Sacks, entitled 'Humphry Davy: Poet of Chemistry', in his *'Everything in its Place: First Loves and Lost Tales'*.[1]

In Sacks's evocative essay, which extends well beyond Davy's scientific accomplishments to his philosophical and religious beliefs, the opening paragraph has especial resonance:

> *'Humphry Davy was for me—as for most boys of my generation with a chemistry set or a lab—a beloved hero; a boy himself in the boyhood of chemistry; an intensely appealing figure, as fresh and alive after a hundred years in his way as anyone we knew. We knew all about his youthful experiments—from nitrous oxide ... to his often reckless experiments with alkali metals, electric batteries, electric fish, explosives. We imagined him as a Byronic young man with wide-set, dreaming eyes ....'*

Albemarle Street: Portraits, Personalities, and Presentations at the Royal Institution. John Meurig Thomas,
Oxford University Press. © Sir John Meurig Thomas 2021.
DOI: 10.1093/oso/9780192898005.003.0003

Sacks then goes on to recall how happy he was growing up as a young scientist in South Kensington, London:

> *'where the history of chemistry, especially its beginnings in the late eighteenth and nineteenth centuries, was laid out; in love, perhaps, most of all, with the Royal Institution, much of which still looked and smelled exactly as it must have when the young Humphry Davy worked there, and where one could browse among and ponder his actual notebooks, manuscripts, lab notes and letters.'*

In this account, we concentrate overwhelmingly on Davy's achievements—as a scientist, in the various places he worked, especially the Royal Institution (RI), and in his public endeavours. Philosophical attributes associated with him are not discussed. We focus primarily on his astonishing range of discoveries.

It was at the RI in Albemarle Street, in the early decades of the nineteenth century, that Davy's pellucid presentations popularized science, especially chemical science. He combined felicity of literary and poetic expression with brilliant scientific discovery and demonstration. He was, arguably, the first-ever popularizer of the physical sciences. And, as mentioned earlier, such was his popularity in London among the intellectuals and aristocrats at the pinnacle of his lecturing fame, that Albemarle Street became the first one-way street in the metropolis, owing to the congestion caused by the numerous carriages that made their way to his performances.

In a literal and metaphoric sense, he was the talk of the town. In George Eliot's famous novel *'Middlemarch'*, both Davy's science and his poetry are described. Furthermore, Jane Haldimand Marcet (see Figure 3.1), the wife of a London-Swiss doctor, constantly attended Davy's lecture-demonstrations, which then became a central and influential feature of her famous 1806 book *'Conversations in Chemistry'*, aimed at young ladies. Marcet inspired Faraday with a love of science in the days before he ever met Davy. Yet, it was Davy that inspired her in the early 1800s. She was born in London of a prosperous Swiss family. At the age of thirty, she married Alexander Marcet, a London physician and chemist, and her husband's circle of friends included the great Swedish chemist J. J. Berzelius, Davy, novelist Maria Edgeworth, and the botanist A. de Candolle.[2] Her book was the most successful elementary chemistry text of the period, especially in America. It was first published in London, but from 1806–30, American publishers made twenty-three impressions of various editions of her work. Her work even outsold, in the US, all the works of Byron. Thomas Jefferson is said to have advised a young student, who asked what he should do about his chemical education, *'read Mrs Marcet's book'*.

It is prudent to outline at the outset of this chapter a summary of Davy's life achievements, many of which were accomplished in Albemarle Street:

(1) He discovered sodium, potassium, boron, calcium, barium, and cadmium, and demonstrated that chlorine and iodine are elements.

Figure 3.1 *Humphry Davy.*
(Courtesy the Royal Society)

(2) He invented the miner's safety lamp, which not only enabled coal miners safely to take a flame (for illumination) into underground seams rich in the explosive gas firedamp (methane), but also signified, quantitatively, the presence of the firedamp from a change in the nature of the flame.

(3) He invented a means for giving light in explosive mixtures of firedamp in coal mines by consuming the firedamp.

(4) He discovered the anaesthetic properties of nitrous oxide (laughing gas), which was used extensively for more than a hundred years in dentistry and surgery; some dental surgeons and anaesthetists still use this gas for special purposes. He also conducted some crucial experiments in physiology (see below).

(5) He invented the technique of cathode protection to arrest the corrosion of materials such as iron and steel when exposed to aqueous solutions

Figure 3.2 *Jane Haldimand Marcet (1769–1858).*
(Wikipedia open access)

or gases rich in water vapour. This he did by placing them in intimate contact with metals such as tin or copper, which corrode sacrificially. (All the ships that sail the oceans of the Earth, and most large buildings, use cathodic protectors to this day.)

(6) He invented the carbon arc, which for many decades was used for street lighting in Paris and other European cities.

(7) He demolished Lavoisier's contention that all acids contain oxygen. So deeply ingrained was Lavoisier's notion of the nature of acids that there is a permanent remnant of it in several European languages. In German, the word for oxygen is *'sauerstoff'*. In Russian it is *'kislorod'*, again meaning 'sour stuff'. From his experiments on hydrochloric acid, he demonstrated that it contained no oxygen. Rather, he showed that all acids contain hydrogen.

(8) He discovered clathrates, which are solids in which open cages of water molecules in a crystalline state encapsulate gaseous species such as chlorine. Nowadays, it is recognized that there is more carbon 'locked up' as methane in clathrates on all the ocean beds and elsewhere on the Earth than all the carbon in fossil deposits.

(9) He pioneered electrochemistry. From the time he read Alessandro Volta's famous memoir (addressed to the president of the Royal Society, Sir Joseph Banks, in 1800), *'On the Generation of Continuous Electricity by the Mere Contact of Two Dissimilar Metals'*, Davy felt that this was the wrong deduction. Volta had interposed sheets of zinc and silver with a conducting substance of another nature, such as discs of paper or cloth that were moistened or preferably soaked in brine, which made them better conductors. Volta built a 'pile' of discs in the order zinc, silver, moistened paper, zinc, silver, and moistened paper.[3] In this way, Volta had made the first-ever electrical battery. But Davy felt that it was not physical contact between dissimilar metals that was the source of the electricity; rather, he believed it was chemical action. In 1800, Davy published five papers in which he demonstrated conclusively that chemical action is indeed the root cause of the generated electricity.

(10) We now realise that, in 1808, Davy first recorded the appearance of a solvated electron.[4]

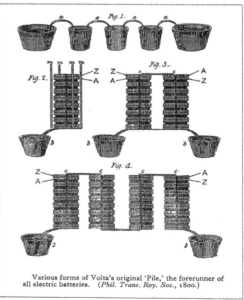

Various forms of Volta's original 'Pile,' the forerunner of all electric batteries. (*Phil. Trans. Roy. Soc.*, 1800.)

Figure 3.3 *The voltaic pile (see text).*

(Courtesy the Royal Society)

These experiments won him great acclaim from Berzelius.[5] And when Davy 'inverted' the argument, i.e., argued that electricity could cause chemical reactions, he proceeded towards the isolation of potassium and, soon thereafter, sodium.

This was what was subsequently called electrolysis, which can be used to break down chemical compounds into their elements.

Davy's Bakerian Lecture[6] to the Royal Society in 1806 on electrochemical considerations was described by Berzelius as *'one of the greatest memoirs in the history of science'*. Such work soon led in 1807–8 to the isolation of potassium, sodium, and calcium. The work won for him the Gold Medal awarded by Emperor Napoleon. And Mendeleev, many years later, expressed the view that Davy's discovery of a way to isolate sodium and potassium (to which we return briefly, below) was one of the greatest discoveries in science.

We next proceed to describe Davy's early years and career, prior to his arrival at Albemarle Street.

3.2 How Did Davy's Life Unfold?

Humphry Davy left school in Cornwall at the age of fifteen. He had been thoroughly educated in the so-called grammar schools of Penzance and Truro, and he was fortunate in the older people who took an interest in him, especially after the death of his father (a woodcarver) in 1794. Equipped with a gun that his father had purchased for him, he would roam the countryside and partake of shooting, hunting, and fishing, activities that he pursued for the rest of his life. He also made a practice of talking to the Cornish miners, many of whom were gifted, often self-taught, practical engineers. He relished going on solitary wanderings and indulging in pantheistic contemplation. An outstanding quality that manifested itself quite early in his teens was the ability to hold an audience. He was a great storyteller, a gift he seems to have inherited from his grandmother. In *'The Collected Works of Sir Humphry Davy, Bart.'*, we read about him as a teenager: *'After reading a few books, I was seized with a desire to narrate, to gratify the passions of my youthful auditors. I gradually began to invent and form stories of my own. Perhaps this passion has produced all my originality. I never had a memory. I never loved to imitate, but always to invent: this has been the case in all the sciences I have studied. Hence many of my errors.'*

He became apprenticed to a surgeon-apothecary (later a distinguished surgeon) who had a fine library, where he read Lavoisier's treatise on chemistry and Nicholson's *'Dictionary of Chemistry'*. On leaving school, he was equipped with a working knowledge of Latin and Greek, and some French and Italian. His initial intention, after his interest in roaming adventures subsided, was to study medicine in Edinburgh. Towards that end, he drew up a prospectus of study when he was barely sixteen years old (Figure 3.3). This scheme of self-education was proposed in 1795.

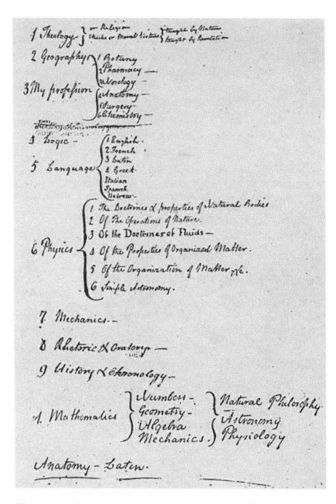

Figure 3.4 *Scheme for self-education proposed by Davy in 1795.*
(Courtesy Ronald King)

While pursuing these studies, he read omnivorously. In March 1798, he came across a paper by an American physician, Dr Samuel Latham Mitchill, who sought to prove that the *'gaseous oxide of azote'* (nitrous oxide) was the principle of contagion, was instantly fatal when breathed, and accounted for the sudden deaths of those stricken by the plague. Highly sceptical of all this, Davy set out to test the hypothesis by experiment, first exposing wounds and the bodies of animals to the gas and then—with his characteristic courage—breathing it himself mixed with air.

He observed no *'remarkable effects'*.

As a result of his friendship with Gregory Watt, the son of James Watt,[7] who had been sent to Cornwall to convalesce, and through his acquaintance with Davies Giddy (later Davies Gilbert, who became President of the Royal Society), Davy's experiments were brought to the attention of Dr Thomas Beddoes once a reader in chemistry at Christ Church, Oxford, and a somewhat eccentric individual.[8] He was also a physician-philosopher, and a well-intentioned philanthropist. He had set up his Pneumatic Institute in Clifton, Bristol, sponsored in part by master potter Thomas Wedgewood and other friends. This institute was part clinic, part lecture centre, and part research laboratory. *Inter alia*, it set out to investigate the curative properties of newly discovered gases—so-called 'factitious airs.' Beddoes needed an assistant, and, impressed by accounts of Davy's experiments, offered the post to him.

At Bristol, opportunities of an entirely new kind presented themselves to the young Davy. Mrs Beddoes (much younger than her husband) was the sister of Maria Edgeworth, who used to visit her frequently and drew to the Beddoes home many of the young writers of the West Country. This is how Davy came to form firm friendships with the poets Samuel Taylor Coleridge and Robert Southey. Through them, Davy formed strong links with William Wordsworth. All these men of letters were fascinated by Davy's wide interests and enthusiasms. Davy showed them some of the poems he had composed in Penzance, and a few of them were published in Southey's '*Annual Anthology*', including one called 'The Sons of Genius' that he had written aged seventeen. Coleridge and Southey were much impressed by Davy.

At the Pneumatic Institute, Davy resumed his experiments with nitrous oxide, which he prepared in copious quantities by the thermal decomposition of ammonium nitrate ($NH_4NO_3 \rightarrow N_2O + 2H_2O$). On inhaling nitrous oxide himself and carefully recording the effects, he drew his own conclusions.[9]

Figure 3.5 *Images of Coleridge, Wordsworth, and Southey. The three Romantic poets befriended Davy.*

(Public domain)

SPECIAL ARTICLES

Anesthesiology, V 114 • No 6   1282   June 2011

Humphry Davy

His Life, Works, and Contribution to Anesthesiology

Nicholas Riegels, M.D.,* Michael J. Richards, B.M., F.R.C.A.†

Davy's work thereby foresaw the ongoing transformation of medicine from a dogmatic, speculative discipline into a rational, experimental science.

Although Davy's work on respiratory physiology and nitrous oxide anesthesia had little practical impact in his own time, he bequeathed to us a foundational legacy of scientific inquiry that endures to this day.

Figure 3.6 *Excerpt from a paper by Nicholas Riegels and Michael J. Richards,* Anaesthesiology, ***2011,*** *114, 1282.*

(Wolters Kluwer Health, Inc., permission granted 14 August 2020)

Davy is reported as saying: '*My sensations were now pleasant. I had a generally diffused warmth…a sense of exhilaration similar to that produced by a small dose of wine and a disposition of muscular motion and merriment.*'

The news of Davy's 'laughing gas' spread, and several of his friends, including the poets Coleridge and Robert Southey, tried it. The effects on patients were examined with scrupulous care and found to be mixed, occasionally beneficial. In his experiments to examine the reaction when nitrous oxide was absorbed in the bloodstream, Davy was the first to devise a means of measuring the residual capacity of the human lung. According to a recent paper[10] (see Figure 3.6) by Nicholas Riegels and Michael J. Richards, Davy measured total lung capacity as 4,700 ml. In this report, Davy's contributions to anaesthesiology are assessed: '*Davy's work thereby foresaw the ongoing transformation of medicine from a dogmatic, speculative discipline into a rational, experimental science. Although Davy's work on respiratory physiology and nitrous oxide anaesthesia had little practical impact in his own time, he bequeathed to us a foundational legacy of scientific inquiry that endures to this day.*'

It is little wonder, therefore, that in the early days of the RI, when Rumford and others contemplated its precise mission (see Chapter 2), Davy was one of those in favour of incorporating medicine as one of the subjects to be pursued.

The results of Davy's year-long experiments using nitrous oxide were published by him in 1800 in a 580-page volume entitled '*Researches, Chemical and Philosophical; Chiefly Concerning Nitrous Oxide, or Dephlogisticated Nitrous Air, and Its Respiration*'. This is a work of admirable orderliness and systematic observation, without speculation, and astonishing accuracy at a time when gas analysis was in its infancy. It established his reputation as an experimentalist of the first order. Although this book attracted much attention, it took some four decades before the significance of one of his observations was fully appreciated and adopted by dental surgeons (in Europe and North America). Davy had stated the following: *'As nitrous oxide in its extensive operation appears capable of destroying physical pain, it may probably be used with advantage during surgical operations in which no great effusion of blood takes place.'*

Even before he left Cornwall for the Pneumatic Institute in Bristol, Davy, as a young man carried out work of a pioneering quality in anaesthesiology. He was, for example, the first to record the medical condition known as laryngospasm, which he discovered in endeavouring to breathe pure carbon dioxide. At an even earlier time he had explored some basic physiological phenomena. At approximately eighteen he had discovered that whereas humans and animals inhaled oxygen and expelled carbon dioxide, plants and trees did the very opposite: they absorbed CO_2 and liberated O_2.[9]

Davy's other activities at the Pneumatic Institute concerned his pioneering work on electrochemistry, prompted, as stated above, by Volta's memoir to Sir Joseph Banks. Among the many insights that he described at that time was his proof that ions move. This is illustrated in Figure 3.7.

The work carried out by Davy in Bristol attracted considerable attention, and, on 31 January 1801, he wrote to his mother telling her that Count Rumford had offered him a post at the RI. In February of that year, the Committee of Managers there decided to offer Davy the posts of assistant lecturer in chemistry, director of the laboratory, and assistant editor of its journals. He was allowed a room in the

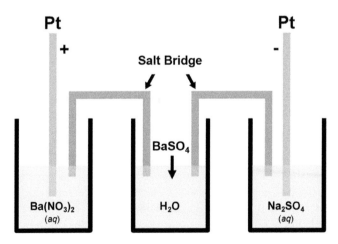

Figure 3.7 *The fact that a precipitate of $BaSO_4$ formed in the middle vessel proved that Ba^{2+} and SO_4^{2-} ions moved in opposite directions under the influence of an applied field.*

house in 21 Albemarle Street, coal and candles, and a salary of one hundred guineas per annum.

3.3   Davy's Years at the RI

The perspicacious and oleaginous Count Rumford knew how to pick winners. Prior to attracting Davy to the RI, he had recruited Thomas Young—whose omnivorous interests and multiple skills we have described earlier in Chapter 1 and will again in Chapter 11—as Professor of Natural Philosophy.

Yet, on arrival at the RI, Davy was soon involved in his own brilliant lectures, which were received with much acclaim. But the demonstrations involving inhalation of nitrous oxide were also the butt of ridicule, not least by the contemporary cartoonist James Gillray.[11] Soon, however, Davy's coruscating lectures attracted huge attention. He was lionized and idolized for the combination of his brilliant lecturing and demonstrating skills. He was also criticized for being self-absorbed. Some members of the public believed him to be ultra-ambitious, arrogant, and not beyond dealing in chicanery and sophistry. It seems that the breadth of adulation blurred the bloom of his youthful innocence and simplicity.

Many of his aphorisms, presented during the rapid flow of his fluent lectures, remain memorable:

> '*Of modern chemistry it may be said that its beginning is pleasure, its progress knowledge, its objects truth and utility. Also, the human mind is always governed, not by what it knows but by what it believes, not by what it is capable of attaining, but by what it desires.*'

Early in the summer of 1801, just three months after his arrival at the RI, Davy wrote to his friend in Bristol, John King, '*The voice of fame is still murmuring in my ears—my mind has been excited by the unexpected plaudits of the multitude—I dream of greatness and utility—I dream of science restoring to nature what luxury, what civilization have stolen from her—pure hearts, the forms of angels, bosoms beautiful and panting with joy and hope—my labours are finished for the season as to public experimenting and public enunciations. My last lecture was on Saturday evening. Nearly 500 persons and unbounded applause. Amen. Tomorrow a party of philosophers meet at the Institution to inhale the joy-inspiring gas—It has produced a great sensation. Ça Ira… I have been nobly treated by the Managers. God bless us. I am about 1,000,000 times as much a being of my own volition as at Bristol. My time is too much at my own disposal. So much for egotism—for weak, glorious, pitiful, sublime conceited egotism.*'

For twelve years Davy was employed at the RI, becoming its Professor of Chemistry in 1802 and its Director in 1804. He retired as Director in 1812; thereafter, he held the post of Honorary Professor up to his death in 1829.

In 1807, he isolated potassium. What, in effect, he did was to reverse the principles that were involved in Volta's discovery. Instead of producing electricity

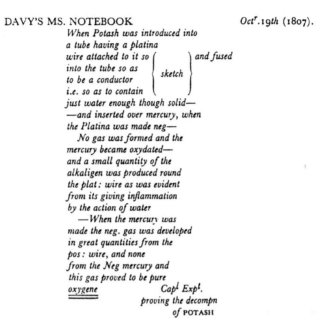

DAVY'S MS. NOTEBOOK   *Oct^r.19th* (1807).

When Potash was introduced into
a tube having a platina
wire attached to it so
into the tube so as   sketch   and fused
to be a conductor
i.e. so as to contain
just water enough though solid—
—and inserted over mercury, when
the Platina was made neg—
   No gas was formed and the
mercury became oxydated—
and a small quantity of the
alkaligen was produced round
the plat: wire as was evident
from its giving inflammation
by the action of water
   —When the mercury was
made the neg. gas was developed
in great quantities from the
pos: wire, and none
from the Neg mercury and
this gas proved to be pure
oxygene   Cap^l Exp^l.
                    proving the decompn
                       of POTASH

Figure 3.8 *Printed version of Davy's laboratory notebook entry for 19 October 1807. He had isolated potassium on 6 October, and he was now anxious to show that potash was the oxide of this new substance. Electrolyzing potash in a closed tube yielded pure oxygen, establishing this to be so.*

(RI—unrestricted permission granted for images used)

by chemical action, he drove a chemical reaction by electricity. He was the first, it seems, to realise this. He electrolyzed potash. A printed version of the page in his laboratory notebook is shown in Figure 3.8 for October 1807.

His note reads: *'Capl Expt. Proving the decomposition of potash'* (from John King, 'Humphry Davy'). (Courtesy RI).

His notebooks make interesting reading. A particular favourite of mine is his entry for 13 September 1807, specifying attributes much wanted in the laboratory of the RI. These included cleanliness, neatness, and regularity. It is amusing that the page itself is heavily smudged!

Davy tended to leave till the last few days or hours the preparation of what he would say in any given lecture; but he rehearsed carefully what he actually did and demonstrated at such events. Figure 3.8 shows the first few sentences of his lecture on electrochemical science on 12 March 1808.

He genuinely did practice what he believed when he said, *'There is no desire more alive and ardent in my mind than that of having it in my power to combine experiments made for the advancement of science with the details of public lectures.'*

Berzelius, the great Swedish scientist, said that his visit to London in the summer of 1812 constituted the most memorable days of his life.[12] At Greenwich, he

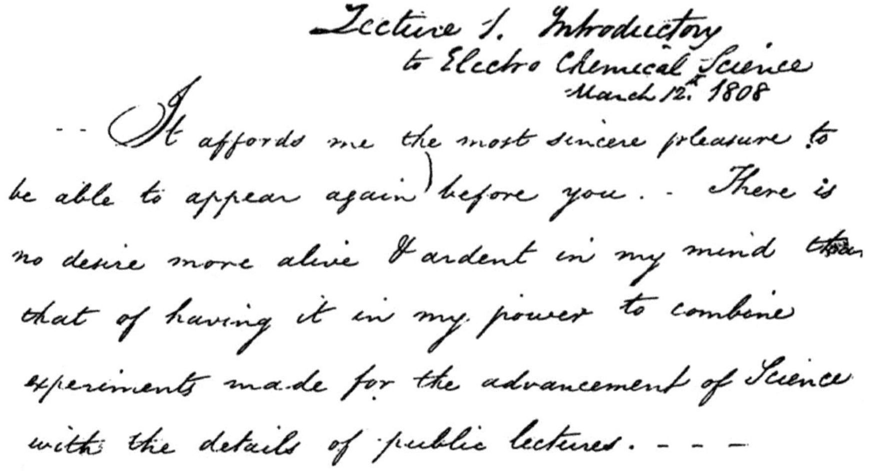

Figure 3.9 *Opening sentences, in his own hand, of Davy's popular lecture at the RI on electro-chemical science, 12 March 1808.*

(Courtesy Mrs Irena McCabe)

met the leaders of the learned world of England, and at dinner he was placed between William Wollaston and Thomas Young and opposite the astronomer Sir William Herschel, the discoverer of the planet Uranus and the man who had demonstrated that 'the rays of the Sun in the prismatic spectrum are much hotter at the red end than at the violet.' He also met Smithson Tennant and James Watt. Wollaston had just discovered palladium and rhodium in 1804 and Tennant osmium and iridium in the same year. (See Figure 3.10.) Berzelius felt completely at home when he visited Davy's laboratory at the RI, of which he wrote: *'Davy took me to his laboratory belonging to the Royal Institution. Although it was fairly well planned, there were so many projects going on that it looked just as spattered and untidy as my own kitchen. There was that kind of untidiness which is unavoidable in any laboratory where one works and where one must have everything easily accessible without having to go to cupboards and shelves each time something is needed. When I saw the collection of broken vessels, of melted slagged retorts, those tables covered with marks from acids and caustic alkalies and ring after ring left by vessels whose contents had boiled over, salt deposits everywhere, badly-handled platinum crucibles, cracked porcelain bowls, all the iron objects rusted by acid vapors, the brass green, pitch-covered articles that had lost half of their japan covering, and files, knives, tongs, valves, etc., lying in a jumble in all the drawers—then I arrived at the happy conviction, which previously had been only a guess, that a tidy laboratory is the sign of a "lazy chemist".'*

Sir William Herschel   William Wollaston

J.J. Berzelius

Thomas Young

Berzelius at the age of 47.
Portrait by Johan Way, 1806. The Library of the R. Swedish Academy of Science.

Figure 3.10   *Photographs of Sir William Herschel, William Wollaston, J. J. Berzelius, Thomas Young, and the cover of Berzelius' book.*[12]

(Courtesy University of Stockholm)

3.4   Davy's Poetic Interests

As mentioned above, Davy in his early life had befriended the English Romantic poets Samuel Taylor Coleridge, Robert Southey, and William Wordsworth. Some of the fascinating consequences of his friendship with the poets are described by R. Holmes.[13] Davy was involved in proofreading the *'Lyrical Ballads'*, published by Wordsworth and Coleridge, and acknowledged by scholars of English literature as a watershed. This collection of poems contains some highly popular pieces, notably 'Tintern Abbey' by Wordsworth and 'The Rime of the Ancient Mariner' by Coleridge. It is obvious from an analysis of Davy's temperament and interests that he could relate in several respects to Wordsworth's sentiments expressed in 'Tintern Abbey'—which Davy expounds in his own writings—and 'Nature Never did Betray the Heart that Loved her'. Other aspects of this great poem that deal with the transformative and restorative qualities of nature also appealed to him.

At the age of seventeen, Davy wrote one of his best-known poems, 'The Sons of Genius', a few verses of which are cited below. This and later poems reveal his pantheistic devotion to nature, his hopes of immortality, and the use of scientific metaphor.[14]

> 'Inspired by her the sons of Genius rise
> Above all earthly thoughts, all vulgar care,
> Wealth, power, and grandeur, they alike despise,
> Enraptured by the good, the great, the fair.'

And,

> 'From these pursuits the sons of Genius scan
> The end of their creation, hence they know,
> The fair, sublime, immortal hopes of man
> From where alone undying pleasures flow.'

Suggestions of his belief in immortality are seen also in another early poem:

> 'And as in sweetest, soundest slumber
> The mind enjoys its happiest dreams
> And in the stillest night we number
> Thousands of worlds in starlight beams
> So we may hope the undying spirit
> In quitting its undying form
> Breaks forth new glory to inherit
> As lightening from the gloomy storm.'

The poetic flair that Davy brought to his science is beautifully illustrated in the opening paragraph of his article 'Some Experiments and Observations on the Colours used in Paintings by the Ancients':[15]

> 'The importance the Greeks attached to pictures, the estimation in which their great painters were held, the high prices paid for their most celebrated productions, and the correlation existing between different states with regard to the possession of them, prove that painting was one of the arts most cultivated in ancient Greece; the mutilated remains of the Greek statues, notwithstanding the efforts of modern artists during three centuries of civilization, are still contemplated as the models of perfection in sculpture, and we have no reason for supposing an inferior degree of excellence in the sister art, amongst a people to whom genius and taste were a kind of birthright, and who possessed a perception, which seemed almost instinctive, of the dignified, the beautiful, and the sublime.
>
> The works of the great masters of Greece are unfortunately entirely lost. They disappeared from their native country during the wars waged by the Romans with the successors of Alexander, and the later Greek republics; and were destroyed either by

*accident, by time, or by barbarian conquerors at the period of the decline and fall of the
Roman Empire.'*

Davy wrote poetry all his life, as did several of his accomplished contemporaries.
According to J. Z. Fallmer,[14] who analysed all aspects of Davy's life: *'Certainly
part of this early poetic output represented the wistful product of an inchoate adolescent
yearning; other of his efforts were "poemes" politely crafted to flatter their recipients, but
still others functioned for Davy as emotional safety-valves, genuine attempts to express
deeply held sentiments which times or circumstances kept him from voicing more openly.'*

3.5 A Selection of Davy's Other Achievements

Davy was elected a Fellow of the Royal Society in 1803 and, in 1807, at the early age
of twenty-nine, he became Secretary of that august body. In 1820, upon the death
of Sir Joseph Banks, who had been President for forty-two years, Davy was elected
President of the Royal Society, a post he held until his untimely death in Geneva in
1829. During his tenure, which coincided with his role as Chairman of the
prestigious London club the Athenaeum, he implemented the policy of weeding out
those amateur-dilettante Fellows, whose knowledge of natural philosophy was rather
tenuous, and 'persuaded' them to instead become members of the Athenaeum.

His role in London public life was important. He was a founding Fellow of the
Zoological Society and the Geographical Society of London. He was also largely
instrumental in founding the Athenaeum in 1823.[16]

Two major achievements of his which will always be associated with his name
were, first, his foundational work in agricultural chemistry, and, of even greater
importance and renown, his invention of the miner's safety lamp.

3.5.1 Agricultural Chemistry

Early in the nineteenth century, there was no well-developed body of knowledge
and technique in agricultural chemistry—as there was, for example, in mineralogy.
When Davy first began his agricultural investigations, Sir Thomas Bernard, who
had been one of the principal founders and patrons of the RI, offered for Davy's
use a piece of ground near his villa in Roehampton. Under Davy's supervision, a
wide range of experiments were carried out there over several years. Davy worked
out the productivity and nutritional value of various grasses and also the nutrients
required to be added to soil to facilitate growth. (This skill made him a favourite
of the aristocratic land-owners of England and Ireland.) He carried out a multitude of experiments and tests, and the book that he published in 1813 (see
Figure 3.10) for the so-called Board of Agriculture, based on the work done by him
in the period 1802–12, contained ninety-seven appendices detailing the yields and
other qualities of various grasses. This book appeared in a total of four editions,
the last in 1827.

ELEMENTS

OF

AGRICULTURAL CHEMISTRY,

IN

A COURSE OF LECTURES

FOR

THE BOARD OF AGRICULTURE.

———

BY

SIR HUMPHRY DAVY, LL. D.

F.R.S. L. & E. M.R.I.

MEMBER OF THE BOARD OF AGRICULTURE, OF THE ROYAL IRISH ACADEMY, OF THE
ACADEMIES OF ST. PETERSBURGH, STOCKHOLM, BERLIN, PHILADELPHIA, &c.;
AND HONORARY PROFESSOR OF CHEMISTRY TO THE ROYAL INSTITUTION.

———

LONDON:

PRINTED BY W. BULMER AND CO. CLEVELAND-ROW ;

FOR LONGMAN, HURST, REES, ORME, AND BROWN,

PATERNOSTER-ROW ;

AND A. CONSTABLE AND CO. EDINBURGH.

1813.

Figure 3.11 *Davy's book on agricultural chemistry.*

(Public domain)

In retrospect, it may be said that although it contained some forgivable errors—for example, an adherence to the humus theory, which held that the source of carbon for plant growth was the humus in the soil—his book was the first significant attempt to present a systematic treatment of the scientific knowledge fundamental to agriculture.

3.6 The Miner's Safety Lamp

Following an explosion at the Brandling Main colliery near Gateshead, on the River Tyne, in which ninety-two men and boys were killed, a society had been set up in October 1813 to investigate the causes and suggest any remedies. It was decided to seek the advice of Davy, who was indulging in his sporting activities in the Highlands of Scotland[17] when the plea for help came. When approached, Davy readily undertook to help, writing *'It will give me great pleasure if my chemical knowledge can be of any use in an inquiry so interesting to humanity.'*

He soon visited the area to see for himself the nature of the problem and he retrieved samples of firedamp from a number of sources. On returning to Albemarle Street, he immediately set to work with Michael Faraday as his assistant on a methodical investigation. He first confirmed the nature of firedamp as methane (CH_4) and proceeded to examine the explosive properties of mixtures with different proportions of air. They discovered that it was most explosive when mixed with seven or eight times its volume of air, and that it was still explosive at a dilution of one in fourteen. Davy also found that firedamp needed a higher temperature to ignite as an ethylene mixture with air. A key experiment was then to examine the expansion of the mixture when it exploded and how the explosion was conveyed through an aperture from one container to another filled with explosive mixture.

Davy's major discovery in all this was that if the communication between the vessels was long enough, and of sufficiently small diameter, the explosion would not pass from the one container to the other. He interpreted this to mean that it was the cooling effect of the connecting tube that was crucial. He demonstrated that the explosion passed more readily through the glass tubes than similar metal ones. He further examined the effect of excess carbon dioxide or nitrogen with the explosive mixture and found that they reduced the tendency to explode.

All this enabled him to set about designing the first safety lamp, a closed lantern to which the entry of air was limited by narrow tubes and in which the chimney was likewise protected. Davy realised that by admitting only a limited supply of air to an oil burner, in a closed lantern, the amount of carbon dioxide and dinitrogen would be sufficient to prevent an explosion if the air were contaminated with firedamp.

It is instructive to quote from Davy's first paper[18] (on his miner's lamp) that he read to the Royal Society on 9 November 1815—just two weeks or so after addressing and solving the major problem that he agreed to investigate. The following is one of Davy's statements:

> *'It is evident, then, that to prevent explosions in coal mines, it is only necessary to use air-tight lanterns supplied with air from tubes or canals of small diameter, or from apertures covered with wire gauze placed below the flame, through which explosives cannot be communicated, and having a chimney at the upper part, on a similar system for carrying off the foul air; and common lanterns may be easily adapted to the purpose, by being made air-tight in the door and sides, by being furnished with the chimney, and the system of safety apertures below and above.'*

Davy's lamps were tested in Wallsend Colliery, by John Buddle, with complete success. Buddle wrote to Davy in the following terms: *'I first tried it in an explosive mixture on the surface, and then took it into a mine...it is impossible for me to express my feelings at the time when I first suspended the lamp in the mine and saw it red hot...I said to those around me: "we have at last subdued this monster"...It is not necessary that I should enlarge upon the national advantages which must necessarily result from an invention calculated to prolong our supply of mineral coal, because I think them obvious to every reflecting mind; but I cannot conclude without expressing my highest sentiments of admiration for those talents which have developed the properties, and controlled the power, of one of the most dangerous elements which human enterprise has hitherto had to encounter.'*

Davy was urged, by Buddle and others, to take out a patent to protect his invention, which, as Buddle said, would yield him a large income. Davy's reply is highly relevant to those scientists—of which there is now a decreasing number in the academic world—who believe that scientific discovery and invention constitute their own reward. What Davy said was: *'My good friend, I never thought of such a thing: my sole object was to serve the cause of humanity; and if I have succeeded, I am amply rewarded in the gratifying reflection of having done so...More wealth could not increase either my fame or my happiness. It might undoubtedly enable me to put four horses to my carriage, but what would it avail me to have it said that Sir Humphry drives his carriage and four?'*

A contemporary invention made by the engineer George Stephenson to avoid explosions in coal mines was, in many respects, similar in design to the lamp designed by Davy. A bitter controversy arose, and the Royal Society found it necessary to issue a statement affirming categorically that Davy had not only discovered independently of all others *'it's principles of the non-communication of explosions through small apertures...but that he also has the sole merit of having first applied it...as a safety-lamp...'*

In due course, Stephenson was exonerated of plagiarism, and, indeed, coal mines in the north of England regularly used his safety lamp, which they called the 'Geordie'.

To express his appreciation of Davy's invention of the safety lamp, Sir Joseph Banks, the President of the Royal Society, had written to Davy declaring that his work would place the Royal Society higher in popular opinion than all other abstruse discoveries (beyond the understanding of ordinary people). For his safety lamp, Davy was accorded the Rumford Prize of the Society; and in 1818, he was made a Baronet, the first scientist to be awarded such an honour.

Davy's safety lamp was used extensively in the coal mines of Europe from Flanders to Russia and beyond. Indeed, in 1825, Tsar Alexander of Russia sent Davy at 21 Albemarle Street a large silver gilt salver and bowl which was, until recently, used by the Director and his wife at dinners given preceding the Friday Evening Discourses.

Following his work on the safety lamp, Davy continued his researches upon flames, making fundamental observations which laid the foundation for the study of combustion as a branch of physico-chemical science. He discovered the catalytic properties of platinum, and in some further developments of the design of his safety lamp he used this catalytic property to produce ample light in explosive mixtures of methane in coal mines.[17,18]

We shall return again (Chapter 11) to Davy's early work, as his experiments are candidates for some of the most beautiful in physics.

3.7   Marriage

The year 1812 was a memorable one for Davy. On 8 April he was knighted by the Prince Regent, and three days later was married to a rich widow, Jane Apreece, while in between gave the final lecture of his last course (as Director) of lectures at the RI. In this year, also, he published his *'Elements of Chemical Philosophy'*, the first part of which elicited great admiration as it was a brilliant sketch of the history of chemistry; Berzelius, in particular, praised it as a masterpiece.

Lady Davy, a distant cousin of novelist Sir Walter Scott, was described by Scott as *'gay, clever and most actively ambitious to play a distinguished part in London society'*, and in the next eight years her new husband somewhat frittered away his creative energies, and instead found himself involved in rounds of social visits, with his long journeys abroad, allowing no time for his scientific explorations. It was said that Davy quite enjoyed the social round, with splendid opportunities for shooting and fishing. The marriage, however, was not a happy one, and, with the passage of time, they spent less and less time together.

His partial separation from the RI—early in 1813 he was appointed Honorary Professor, a title he held until 1823—did not halt his chemical investigation, for he still had access to the laboratories at 21 Albemarle Street, and he constructed a portable laboratory which he took with him on his travels. A few months prior to his departure in October 1813 for an extended trip on the continent, he (and the RI) hired Michael Faraday as his assistant. Faraday, Lady Davy and her maid, and the portable laboratory all accompanied Davy to the continent.

Davy's circle of friends was remarkable. A trusted adviser of the highest society, he was welcomed as an honoured guest at the great country houses of

England. A painting that now hangs in the Tate Gallery, London, shows Davy present, along with other members of the illuminati, at Thomas Coke's annual sheep-shearing at Halkham. He stayed with the Duke of Bedford at Woburn, with Lord Sheffield in Sussex, and with Lord Byron in Ravenna. Some of his Italian acquaintances included Canova, the Grand Duke of Tuscany, and Count Bardi and Signior Gazzani, who were respectively the Director of the Lincei Academy in Rome and Professor of Chemistry at the Florentine Museum. He knew an even larger number of savants in Paris, where he was admired by most of them.

Increasingly, during his last years of failing health, he journeyed without Lady Davy, she finding her satisfactions in the social life of London, and he his consolations in travel. In March 1828, he set off for the continent to spend the summer in his favourite haunts in Austria and the winter in Italy, where he experimented with the electric fish—the torpedo. On these journeys, he wrote *'Consolations in Travel'* or *'The Last Days of a Philosopher'*, which would be published posthumously in 1830. In 1829, during a visit to Rome, he suffered a stroke, and on 29 May of that year died in Geneva while attempting to return to England. When his last days approached and he gently hinted to Lady Davy that she come to him, she did so without delay. His brother John came also, and both were with him when he died. He had earlier stated that he wished *'to be buried where I die: natura curat suas reliquias'* ('nature takes care of her own remains'). He is buried in the Cemetary of Plain-Palais, Geneva.

In his notebook as a youth of seventeen in Penzance, Davy wrote: *'I have neither riches, nor power, nor birth to recommend me. Yet if I live I trust I shall not be of less service to mankind and my friends than had I been born with these advantages.'*

REFERENCES

1. O. Sacks, *'Everything in its Place: First Loves and Last Tales'*, Knopf, Canada, **2019**, p. 20.
2. M. S. Lindee, *Isis*, **1991**, *82*, 8.
3. It is an interesting etymological fact that the French expression for battery is *'la pile'*, in recognition of the pile Volta constructed as a source of continuous electricity by placing decks of silver and copper, interspersed alternatively with pads saturated with brine, on top of one another.
4. Peter P. Edwards, 'The Electronic Properties of Metal Solutions in Liquid Ammonia and Related Solvents', in H. J. Emeleus, ed., *'Advances in Inorganic Chemistry and Radiochemistry'*, XXV, A. G. Sharpe-Elsevier, Academic Press, **1982**.
5. Jöns Jacob Berzelius (1779–1848) was a meticulous experimenter and systemizer of chemistry; he also had a flair for coining words for phenomena and substances—'catalysis', 'protein', and 'isomerism' were all introduced by him. He compiled a table of the atomic weights of the elements. He discovered the elements cerium, thorium, selenium, and silicon. Like Lavoisier, he believed in the importance of oxygen. He argued for several years that chlorine contained oxygen.

6. The Bakerian Lecture is the Royal Society's premier lecture in the physical sciences. Instituted in 1775, it calls *'for an oration or discourse to be spoken or read yearly by one of the Fellows of the Society on such parts of natural history or experimental philosophy...on such time as the President and Council shall be pleased to order and support.'* Davy gave the Bakerian Lecture seven times.

7. The famous Scottish inventor and instrument maker James Watt (1736–1819), renowned for the Watt engine, which, throughout its various stages of improvement, was one of the main contributions to the Industrial Revolution.

8. Beddoes was dismissed from Oxford, partly because he was ardently in favour of the French Revolution, a sentiment that distressed his conservative colleagues.

9. For detailed accounts of the experiments that Davy tried upon himself, including drinking a bottle of red wine while under the influence of laughing gas and recording his pulse and other physiological facts, see R. Holmes, *'The Age of Wonder'*, Harper Collins, London, **2016**.

10. N. Riegels and M. J. Richards, *Anaesthesiology*, **2011**, *114*, 1282.

11. James Gillray (1756–1815) was an English caricaturist famous for his etched political and social satires, mainly published between 1792–1810.

12. J. Erik Jorpes, 'Bidrag Till Kungl. Svenska Vetenskaps Akademiens Historia VII', in *'JAC BERZELIUS, His Life and Work'*, Regia Academia Scientiarum Suecica, Stockholm, **1966**.

13. R. Holmes, *'The Age of Wonder'*, Harper Collins, London, **2016**.

14. A full analysis of Davy's poetic qualities is given in J. Z. Fullmer, *'Young Humphry Davy: The Making of an Experimental Chemist'*, American Philosophical Society, Philadelphia, **2000**.

15. Published in *Phil. Trans. R. Soc.*, **1815**, *A106*, 1–22.

16. See the charming account by George Porter of the interaction between the Royal Society, the RI, and the Athenaeum in *'Armchair Athenians: Essays from the Athenaeum'*, London, The Athenaeum, **2001**, p. 95.

17. J. M. Thomas, 'Sir Humphry Davy and the Coal Miners of the World: a Commentary on Davy (1816). An Account of an Invention for Giving Light in Explosive Mixtures of Firedamp in Coalmines', *Phil. Trans. R. Soc. A*, **2015**, *373*, 20, 140, 288.

18. H. Davy, *Phil. Trans. R. Soc. Lond., A*, **1816**, *106*, 23–24.

4

Michael Faraday: Paragon

4.1 Introduction

Michael Faraday's reputation as a scientist and natural philosopher is so elevated and so well known that it is almost supererogatory to attempt yet another description of it. Both Rutherford, the most supremely skilled of experimentalists, and Einstein, who among theoreticians occupied the pinnacle, each sang his praises to high heaven. Faraday's name and achievements are evergreen and an endless source of interest, enlightenment, and inspiration. All commentators of his, including historians of science, no less than most practising experimental and theoretical scientists—even cosmologists—regard his successes and ascent from lowly beginnings to the most exalted of status among his contemporaries as an unending source of fascination.

Moralists, and those who engage in the study of the interface between science and religion, as well as resolute non-believers, praise his irreproachable integrity and intellectual honesty. Consider the manner in which Faraday ends his 122-page paper in *Phil. Trans. R. Soc.,*[1] in 1850, entitled 'Experimental Researches in Electricity, 24th–27th Series. 30. On the Possible Relation of Gravity to Electricity':

'Here end my trials for the present. The results are negative; they do not shake my strong feeling of an existence of a relation between gravity and electricity, though they give no proof that such a relation exists.'

Fresh students of his experimental and public work in disparate fields, such as organic chemistry, catalysis, metallurgy, colloid science, optics, liquefaction of gases, photochemistry, and the popularization of science (which do not cover all of his activities), are often stimulated on discovering the novelty of his experimental approach. This is particularly true of the ingenious experiment when Faraday discovered the paramagnetism of gaseous oxygen by encapsulating it in a soap bubble and passing it through the poles of a magnet. They are also impressed by his pertinacity and judgement in recording in a sequence of numbered paragraphs in his notebooks—there are 16,041 of them between 1831 and 1862—his immediate thoughts on the meaning, nature, and possible consequences of his experimental observations.

Albemarle Street: Portraits, Personalities, and Presentations at the Royal Institution. John Meurig Thomas, Oxford University Press. © Sir John Meurig Thomas 2021.
DOI: 10.1093/oso/9780192898005.003.0004

Since many accounts of Faraday's life and of his vast contributions as a scientist are available,[2-13] this chapter will focus only on some aspects of Faraday's work and behaviour that made him such a paragon as a natural philosopher and a human being. The Appendix of this chapter repeats some of the remarks made by the then Archbishop of York—himself a one-time expert physiologist—at the Bicentenary Celebration of Faraday's birth, which I organized at Westminster Abbey in September 1991. Other aspects of Faraday's unique skills and status will emerge in the remainder of this chapter.

As a backdrop to our discussion, Table 4.1, compiled by a prominent student of Faraday's work and influence, Frank James,[9] has produced a useful chronology of some events of Faraday's life. (There are certain omissions here; for example, Faraday's discovery of superionic conductors in solids like PbF_2 and semiconductivity in Ag_2S. It also does not mention the discovery of the synonymity of matter and electricity. A full account of essentially all his numerous discoveries is given in ref [7]).

Table 4.1 *Michael Faraday (1791–1867). A chronology of some events in his life. See also Section 4.4 below for more details*

Date	Details
22 September 1791	Born at Newington Butts, London
7 October 1805	Apprenticed as a bookbinder to George Riebau
30 October 1810	Father dies (19)†
1810–13	Attends meetings of the City Philosophical Society
February–April 1812	Attends Humphry Davy's lectures at the Royal Institution (RI)
7 October 1812	Apprenticeship expires; commences career as a journeyman
December 1812	Interviewed by Davy
1 March 1813	Appointed laboratory assistant in RI
13 October 1813	Leaves on continental tour with Davy
17 April 1815	Returns from continental tour
15 May 1815	Reappointed laboratory assistant in RI
1815–18	Attends meetings of the City Philosophical Society where he gives his first lectures
1818–22	Works on improving steel
21 May 1821	Appointed Superintendent of House of the RI
2 June 1821	Marries Sarah Barnard
15 July 1821	Makes confession of faith in Sandemanian Church
3 September 1821	Discovers electromagnetic rotation
6 March 1823	Liquefies chlorine
8 January 1824	Elected Fellow of the Royal Society
February–May 1824	Founder Secretary of Athenaeum Club

(Continued)

Table 4.1 Continued

Date	Details
1824–30	Works for joint Royal Society and Board of Longitude committee to improve optical glass
7 February 1825	Appointed Director of the Laboratory at the RI
1825	Initiates Friday Evening Discourses at the RI
May 1825	Disovers bicarburet of hydrogen (later renamed benzene by Eilhard Mitscherlich)
1826	Initiates Christmas Lectures for children at the RI
1827	Publishes '*Chemical Manipulation*'
1829	Appointed Scientific Adviser to the Admiralty
1830–51	Lecutres on chemistry at Royal Military Academy, Woolwich
29 August 1831	Discovers electromagnetic induction (40)
1832	Receives Doctor of Civil Law degree from Oxford University
1832	Confirms identify of electricities
1 July 1832	Appointed Deacon in Sandemanian Church
1832–4	Works on electrochemistry and invents, with William Whewell, its nomenclature
18 February 1833	Appointed first Fullerian Professor of Chemistry at the RI
1835–6	Controversy over award of Civil List pension
1835–65	Scientific Adviser to Trinity House
1836	Invents Faraday Cage and explores the nature of electricity
1837	Works on induction
20 March 1838	Mother dies
1839–43	Partially retires from lecturing and research due to ill health
1839	Volume I of '*Experimental Researches in Electricity*' published (48)
15 October 1840	Appointed Elder of Sandemanian Church
1844	Volume II of '*Experimental Researches in Electricity*' published
19 January 1844	Lecture on nature of matter
31 March 1844	Excluded from Sandemanian Church
5 May 1844	Restored to Sandemanian Church
1845	Reports on the Haswell Colliery explosion
13 September 1845	Discovers magneto-optical effect
4 November 1845	Discovers diamagnetism
1845–55	Develops theory of electromagnetic field
3 April 1846	Lecture on ray vibrations
1849	Works on relation of gravity and electricity (58)
6 May 1854	Lecture on mental education

Date	Details
1855	Volume III of '*Experimental Researches in Electricity*' published
1856	Works on transmission of light through solutions
1858	Declines Presidency of the Royal Society (67)
1858–67	Occupies grace-and-favour house at Hampton Court
1859	Publishes '*Experimental Researches in Chemistry and Physics*'
21 October 1860	Appointed Elder of Sandemanian Church
1862	Receives Doctor of Civil Law degree from Cambridge University
20 June 1862	Gives last lecture at RI
1864	Declines Presidency of the RI
5 June 1864	Resigns as Elder in Sandemanian Church
25 August 1867	Dies at Hampton Court (76)
30 August 1867	Buried at Highgate Cemetery

(*Compiled by Frank A. J. L. James*)[9]

†Numbers in brackets refer to Faraday's age.

4.2   A Few of Faraday's Greatest Discoveries

Such has been the impact of Faraday's work on scientific and technological development that one is all too tempted to fall into the common error of saying: '*But for Faraday, we should have been without this or that amenity or scientific development.*' Such a statement, however, can never truly hold good for any scientist. The intricate fabric of scientific progress is such that were a particular discovery not made by one scientist it would inevitably be made, possibly by a different route, by another. (Had X-rays not been discovered by Roentgen in 1895, they would certainly have been discovered in due course by another.)

4.2.1   Electromagnetic Induction

The description of what Faraday discovered on 29 August 1831,[10] given in a charming booklet published by the RI in 1973 (but no longer extant)[11] on Michael Faraday, and composed by my compatriot and former colleague at the RI Professor Ronald King, merits repetition here: '*There appears in Faraday's laboratory notes a most significant entry, headed, "Experiments on the Production of Electricity from Magnetism, etc., etc.", it describes how he had taken a soft iron ring, $\frac{7}{8}$ thick and 6 inches in diameter, and on one half had wound many coils of copper wire, each turn separated from the next by twine and each layer from the next by calico.*

'*Similarly on the other half he had wound another coil separated from the first* (Figure 4.1). *He connected the ends of one coil to a galvanometer—in this experiment simply a wire passing over a magnetic needle as in Oersted's experiment. He then connected the ends of the other coil to a voltaic battery—"immediately a sensible effect on the needle. It oscillated and settled at last in original position. On breaking connection of A side with Battery again a disturbance of the needle." At last an effect—but not quite what he had expected. A current was produced in the second coil only when the current in the first was switched on or off. There was no effect on the second coil when the current in the first was flowing steadily. Here was the key to it all—a changing current in one coil produced a current in the other.*

'*Working from a more sensitive form of galvanometer, he observed that the deflection of the magnet on breaking the current was opposite to that on making. He tried coils of different shapes and sizes and found that he could dispense with the ring altogether: two coils on a cardboard cylinder showed the effect, but it was greater with an iron core.*

'*On 24th September he found a new effect. Two flat magnets were put together North pole to South pole, South pole to North pole* (Figure 4.2(A)). *A coil was wound on a short iron core, the magnets were separated at one end and the core inserted between the poles. The coil was connected to the galvanometer. When one of the magnetic poles was pulled away from the core a current passed through the coil. When it was returned there was a current in the reverse direction. Faraday noted "Hence here distinct conversion of magnetism into electricity".*

'*In December he set to work to produce currents, using only the earth's magnetism. In this he was successful with a variety of arrangements—a rotating disc, a coil of wire with an iron core simply twisted in the hand and, most beautiful of all, a single rectangle of wire turned about an axis perpendicular to the magnetic meridian.*' (See Figure 4.2.)

Fig. 43.7. The original Faraday ring. One of the treasures of the Royal Institution

Figure 4.1 *Faraday's ring*[11] *(photograph of the original).*

(Courtesy Royal Institution (RI))

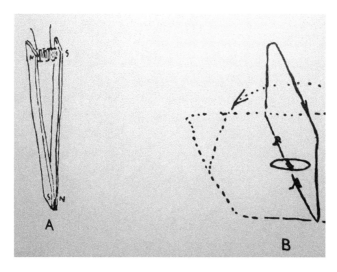

Figure 4.2 *Sketches from Faraday's diary: (A) arrangement of magnets and coil (24 September 1831); (B) a single loop of wire rotated in the Earth's field produces a current (26 December 1826). (Courtesy RI)*

Thus, arguably, the most famous of all Faraday's discoveries, electromagnetic induction, can now be illustrated very vividly, as shown in Figure 4.3, below.[12]

This discovery led to the dynamo, the transformer, and other electrical machines, as well as, later, the industrial-scale generation of energy (and the establishment of electrical engineering as a separate discipline). What is not widely appreciated among members of the public and among some scientists is that numerous instruments deployed in chemical and medical research, such as nuclear magnetic resonance and magnetic resonance imaging (see Chapter 8, Section 8.2), are reliant on electromagnetic induction.

4.2.2 The Laws of Electrolysis

The laws of electrolysis rank among the most accurate generalizations in science. They describe in quantitative terms the relationship between the extent of chemical decomposition of a conducting substance and the amount of electricity that passes through it. The first law of electrolysis in Faraday's words states: *'Chemical action or decomposing power is exactly proportional to the quantity of electricity which passes.'* The second law of electrolysis, again in his own words, asserts: *'Electrochemical equivalents coincide and are the same with ordinary chemical equivalents.'* In other words, the electrochemical equivalent of an element is proportional to its ordinary chemical equivalent. We call the amount of electricity necessary to liberate one equivalent (i.e., 1.008 g of hydrogen and 35.457 g of chlorine) from a solution of hydrogen chloride, or, in short, 1 g equivalent of any element from their conducting

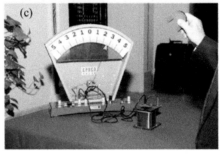

Figure 4.3 *An illustration of the nature of electromagnetic induction in which electricity is produced when conducting wires move past Faraday's lines of force. The energy that goes into the electrical current comes from the work done by moving the magnet. The current passes in one direction when the magnet is inserted into the coil and in the other when it is retracted.*

(I am grateful to Professors J. Ragai and H. Omar, American University in Cairo, for help with this illustration.)

solutions (or conducting molten salts of its compounds), one faraday (which is 96,493 Coulombs of electricity).

These laws brought forth order where there was hitherto confusion. The key factors were simply the quantity of electricity and the chemical equivalents. The significance of this fact is that, fundamentally, chemical forces and electrical forces are intimately and quantitatively related.

As the German physiologist-turned-physicist Hermann von Helmholtz (during the course of his Faraday Lecture at the RI in 1881) concluded: '*Electricity must have a unitary structure, the maximum value of the unit being that which is sufficient to react with one univalent atom.*' Commenting on Faraday's laws, Richard Feynman,[13] as stated in the Preface, said that Faraday had found that the '*atoms of matter are in some ways endowed or associated with electrical powers to which they owe their most striking qualities, amongst them their mutual chemical affinity. He had discovered that the amount of electricity necessary to perform electrolysis of chemical substances is proportional to the number of atoms which are separated, divided by the valence.*'[13]

These discoveries by Faraday led Feynman to say that they constituted *'the most dramatic moment in the history of science'*. To put a gloss on it, Feynman further declared that the result of the combination of iron and oxygen atoms, which make iron oxide, is that some of them are electrically plus and some of them electrically minus, and they attract each other in definite proportions.

4.2.3   Practical and Societal Consequences of Faraday's Laws of Electrolysis[7]

I have dealt with these in a prior publication.[7] Briefly, because of Faraday's laws, electroplating, electro-gilding, and silvering became industrially viable procedures. These commercial developments signalled the rather rapid demise of Old Sheffield Plate (now prized for its antique value), which was produced by heating and annealing sheet silver on to a copper-rich substructure. It could be argued that Faraday's scientific endeavours were the most important factors in shifting world dominance in the silver-plating industry from Sheffield to Birmingham.

Against the walls in the Chapel of Aston Hall, Birmingham, there stands what is claimed to be the first electroplating machine. It contains the following statement: *'To Birmingham belongs the honour not only of introducing electro-plate, the use of which has extended to every civilised nation, but also the honour of first adopting Faraday's great discovery of obtaining electricity from magnetism–a discovery that has influenced science and art to an enormous extent.'*[14]

4.2.4   A Dramatic Lecture-Demonstration

In 1836, Faraday pioneered work on electrostatics and performed his renowned cage experiment in the lecture theatre of the RI. He had earlier come to the conclusion that there was no such thing as an absolute charge; whenever a body is charged, an equal charge of opposite sign is induced on neighbouring bodies. Moreover, Faraday deduced (as had Henry Cavendish long before, but left unpublished) that the charges resided on the surface of a conductor. In his lecture-demonstration, he sat inside a twelve-foot-cubed metallic structure (that he had built) covered in fine (conducting) wire mesh, one side of which had a door through which he could step inside. With the cage insulated from earth, it was charged, via an external machine (a forerunner of the Wimshurst machine), to a potential of approximately 150,000 volts. This caused large sparks and flashes (like artificial lightning) at the outside of the cage. But, Faraday, holding a sensitive electrometer, was unperturbed and unaffected by this fierce electrical activity. This demonstrated, in a daring fashion, that an electrified body carries its charge on the outside surface, a fact that is salutary and reassuring to remember when we fly as passengers in a jet aircraft through storms and lightning. This is the experiment that led to the Faraday Cage.

4.3   Faraday and Franklin: Parallels in their Work

In contemplating the careers of Benjamin Franklin and Faraday, one encounters the confluence of many common, and a number of contrasting, characteristics. Franklin and Faraday, at different times, were each the best known and most admired of men in the Western world: Franklin during the last half of the eighteenth century; Faraday, who was born eighteen months after Franklin's death, during the middle half of the nineteenth century.

Each discovered a large variety of new phenomena, and each was associated with some of the most spectacular (and dangerous) experiments ever performed: the fabled kite experiment in the case of Franklin (which demonstrated that the source of lightning is electrical); the electrified cage (which established that a conducting body is charged on the outside) in the case of Faraday. Electricity was of central importance in their scientific endeavours: its nature, its creation, its control, its utilization. Each in their different ways, and in different contexts, established a language of electrical discourse. The terms 'battery'. 'positive electricity', and 'negative electricity', and the immensely important principle of electro-neutrality, all came from Franklin; the words 'electrolysis', 'electrolyte', 'anode', 'cathode', and 'ion', and the demonstration that matter and electricity are inextricably connected, came from Faraday.

Both Franklin and Faraday achieved mastery over nature without outward prompting—they were driven by some compelling, ineffable, intrinsic curiosity. Both were autodidacts, pursuing the ideas of self-learning and self-improvement. They each believed passionately in prudence, industry, scepticism, intellectual honesty, and the sacrosanctity of evidence. And their accomplishments afforded proof that being highly cultured, and even learned, does not necessarily imply formal education.

Both Franklin and Faraday had an extremely wide range of scientific interests, and their creative intelligence and versatility were exceptional. The scientific books that they wrote were hugely popular—and, incidentally, the title of their most famous scientific texts were uncannily similar:

> Franklin, '*Experiments and Observations of Electricity*';
> Faraday, '*Experimental Researches in Electricity*'.

Like Franklin, Faraday wrote with mellifluous charm and candour. Faraday was the apotheosis of the self-critical scientist. He was a deeply religious man; his favourite verse in the Bible occurs in the Book of Job 9: *'If I justify myself, my own mouth shall condemn me; if I say I am perfect, it shall also make me perverse.'*

He was generous and encouraging to others, but ferocious in the criticism of his own work. He would probably have agreed with Jane Austen, who, in one of her letters, uttered this prayer: *'Incline us, oh God, to think humbly of ourselves; to be severe only in the examination of our own conduct; to consider our fellow creatures with kindness.'*

Despite their towering achievements and iconic status, both Faraday and Franklin had rather poor elementary schooling. Everyone knows that Franklin left school at ten, and at the age of twelve was indentured to a printer. Faraday, when he was thirteen, became an apprentice and errand boy to a London book-binder and bookseller, which provided him with the opportunity of reading widely.

Franklin became a rich businessman, promoting the establishment of such public services as a fire department, a lending library, and an academy, and he was the founder of the American Philosophical Society. Faraday lacked the public, civil, political, diplomatic, and legislative skill of Franklin. In contrast to the affable, humorous, witty, effulgent, and gregarious Franklin, who was not averse to indulging in amorous flirtatious talk (and action), Faraday was of a retiring, almost reclusive nature, and rather shunned socializing. Yet, in his day, especially from the 1830s onward—after his sensational discovery of electromagnetic induction—he was a leading figure in Victorian England. Prince Albert, Victoria's regent, befriended him and attended many of his lectures.

Nowadays, lightning is a phenomenon that will forever be associated with the name of Benjamin Franklin. And the lightning rod of Franklin has had incalcul-able beneficial consequences. For example, the Campanile in Venice has been standing for more than a millennium. But early records refer to several fires and particular destruction.[4] The structure was struck by lightning in 1388, in 1417, and again in 1489. Lightning again damaged the tower severely in 1548, in 1565, in 1653, and in 1745, when it was almost destroyed. Further damage was sus-tained in thunderstorms in 1761 and 1762. In 1766, however, a Franklin lightning rod was installed, and the Campanile had a more comfortable existence until 1902, when a major catastrophe occurred with a fall of thirteen-thousand tonnes of masonry. But, it was not any inadequacy on the part of the Franklin rod that caused this incident; rather, it was the feeble tensile strength of masonry that was the root cause of the destruction.

The trajectory of Franklin's life is so well known that it would be unnecessary for me to expatiate upon it further here. Instead, I shall dwell henceforth on Michael Faraday, the blacksmith's son, born in London, who revered Franklin and frequently quoted Franklin's work, especially the famous apophthegm in response to the question that all scientists who achieve notable distinction encounter: *'Of what use is this discovery?'* To which Franklin, later echoed by Faraday, replies, *'Of what use is a new born baby?'*

4.4   How Did Faraday Get to the RI?

Faraday worked in the bookbinder's shop (Figure 4.4) until he was twenty-one years of age, when he was given tickets by a kindly customer, in the spring of 1812, to go and listen to and observe a performance by the most brilliant star in the European scientific firmament, Sir Humphry Davy.

Figure 4.4 *The apprentice Michael Faraday in Mr Riebau's bookbinder's shop, with a kindly customer who gave Faraday tickets to attend Davy's lectures at the RI.*

(Author unknown)

The thirty-four-year-old Humphry Davy, as we saw earlier, was already famous worldwide by the time that the twenty-one-year-old Faraday saw and heard him. He had discovered sodium, potassium, calcium, and other elements and phenomena described in Chapter 3.

Davy's breathtaking and daring lecture-demonstrations drew the applause of the multitudes. Michael Faraday sat entranced in the theatre of the RI— mesmerized by the pellucidity of Davy's presentations. Faraday took copious notes. And shortly thereafter he rewrote them, included various illustrations and sketches and a detailed index to all four lectures, and bound them impeccably. In

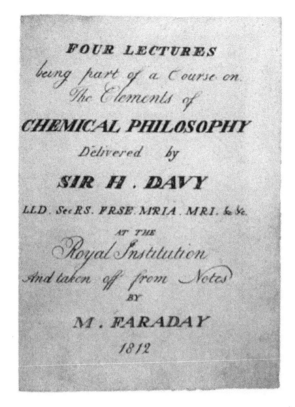

FOUR LECTURES
being part of a Course on
The Elements of
CHEMICAL PHILOSOPHY
Delivered by
SIR H . DAVY
LLD . Sec RS. FRSE. MRIA . MRI. &c.
AT THE
Royal Institution
And taken off from Notes
BY
M . FARADAY
1812

Figure 4.5 *Title page of Faraday's notes on Davy's lectures, 1812.*
(Courtesy RI)

due course, he sent them to Davy (Figure 4.5), along with his letter asking if Davy could offer him a job. Davy replied on Christmas Eve 1812 (Figure 4.6). Faraday, an inveterate collector, kept that letter.

Davy's reply merits comment. The book (of his own lectures) that Faraday had sent him was a document of Mozartian perfection. And Davy's reply began with the words *'I am far from displeased by the evidence you gave me of your power of memory...'* and then went on to say that he would be out of town till late January 1813. The opening words of Davy's letter, however, constitute a classic example of English understatement. Here was Davy receiving a beautiful version of his own lectures, and yet—as only an Englishman would—he says *'I am far from displeased'*! (No Welshman, Scot, or Irishman would have answered in such restrained terms.)

Davy, in due course, interviewed Faraday, who started work, essentially as a bottle-washer, in the laboratories of the RI in March 1813 (on St David's Day). Soon—such was Faraday's dexterity and precocity—he was entrusted with the preparation of samples of the newly discovered nitrogen trichloride, a capriciously

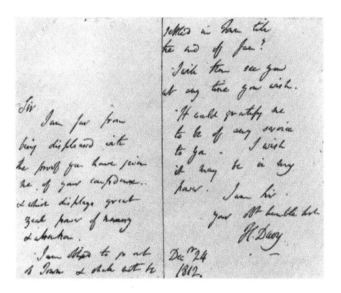

Figure 4.6 *Davy's reply to Faraday (on Christmas Eve 1812).*
(Courtesy RI)

explosive substance. He also assisted Davy in the construction of the miner's safety lamp, an invention that further enhanced Davy's fame.

Starting in October 1813, Michael Faraday accompanied Davy and his wife (and a mobile laboratory) on an extended European tour lasting almost eighteen months. It took them to Paris, Montpellier, Milan, Genoa, Turin, Florence, Geneva, and many other European cities. During the course of the tour they met André-Marie Ampère, François Arago, Pierre-Simon Laplace, Joseph-Louis Gay-Lussac, and Alessandro Volta. All the while, Davy and Faraday conducted experiments: they isolated and identified iodine, for example, while in Paris; in Geneva, they studied the electrical discharge from the torpedo fish; and then in Florence, where they had access to the giant lens of the Duke of Tuscany (that was held in the Accademia Del Cimento), they conducted an important experiment of burning diamond in a vessel containing only oxygen using the focused rays of the Sun to attain the high temperature. As Davy surmised, all the products of the combustion of the last experiment were the oxides of carbon, thereby proving that diamond (like graphite) was pure carbon. It is believed that this experiment was carried out close to, or possibly in, the Boboli Gardens (Figure 4.7).

Many scientists at that time and even later, including James Dewar, a century later, disputed whether diamond was pure carbon.

Davy and Faraday made the acquaintance of Volta in Milan, where Faraday wrote notes on the firefly and the glowworm. They collected the natural flammable gas at Pietra Mala, and identified it as methane (first discovered by Volta).

And on a daily basis, Faraday received expert tuition from Davy, while also acquiring during his peregrinations a working knowledge of French and Italian.

Alessandro Volta

Gay-Lussac and Jean-
Baptiste Biot, 1804

Laplace   Ampère

The complete combustion of diamond in oxygen
using the Duke of Tuscany's lens.

Figure 4.7 *Five of the savants that Faraday and Davy met in their continental tour. On the top left, Gay-Lussac and Jean-Baptiste Biot are shown several thousand feet above Paris in a hydrogen balloon in 1804. It is now believed that the diamond combustion experiment carried out in Florence took place close to the present Boboli Gardens (see text).*

(The illustration of the Gay-Lussac and Biot flight was provided by Professor Sherry Rowlands.)

Paris was brimming with scientific experiments at that time, such as the flight in a hydrogen balloon made in 1804 by Gay-Lussac and Jean-Baptiste Biot.

They returned in May 1815, and thereafter Faraday's career as a natural philosopher ascended rectilinearly to stratospheric altitudes (see Table 4.1). He liquefied, for the first time, some twenty gases (including ammonia, the basis of early and later refrigeration). He discovered and established the chemical formula of benzene, which he prepared by distillation of fish oil (see Section 4.8.1). He invented the first electric motor in 1821. He pioneered organic photochemistry, that is he harnessed sunlight to synthesize new organic compounds. He became a superb analytical chemist (and could have accumulated substantial riches if he had continued to serve as expert witness in legal disputes). He identified isomers of chemical compounds, showing that isobutylene (2-methyl propene), which he was the first to prepare, has an empirical chemical formula exactly the same as that of ethylene (CH_2). He prepared the sulfonic acid derivatives of naphthalene, the precursors of industrial dyestuffs. He improved the optical quality of glass, and it was he who first drew glass fibres of a kind that later were utilized as light guides. He prepared 'stainless' alloy cut-throat razors, made of iron and platinum. He pioneered the study of dielectrics; and such was the magnitude of his contribution to this field that the unit of capacitance is named a farad in his honour. In 1829, he wrote a masterly text on '*Chemical Manipulation*'. He studied heterogeneous catalysis, colloidal metals, and ionic conductivity in inorganic solids such as PbF_2, during which he was the first to note the now topical subject of superionic conductivity. He also identified the phenomenon of semiconductivity, and what is now termed thermistor action (well over a century before it became the centrepiece of electrical and electronic circuitry). There were numerous other discoveries. His successor as Director of the RI, John Tyndall, set out in a series of Discourses in 1868, later published as a book, the astonishing chronicle of '*Faraday as a Discoverer*'. In making the discovery of electromagnetic induction, Faraday argued that a magnetic field surrounded a magnet (Figure 4.8), just as he later argued that a gravitational field surrounds every solid object. In fact, Faraday was the founder of field theory: all theoreticians and cosmologists ever since have acknowledged this fact. Faraday, in his mind's eye, could picture lines of force emanating from a magnet, and he illustrated the reality of this picture by sprinkling iron filings on a paper beneath which he placed a magnet. His lines of force ushered a new era into physics and cosmology; an era built on the concept of field, which pervades the space around a magnet and the Earth—see the remarks by James Clerk Maxwell (much later): '*weaves a web through the sky*'.

In his famous '*A Treatise on Electricity and Magnetism*', James Clerk Maxwell (whose Research Fellowship submission to Trinity College, Cambridge, chose the title '*Faraday's Lines of Force*'), made the following statement:

'Faraday, in his mind's eye, saw lines of force traversing all space where the mathematicians saw centres of force attracting at a distance: Faraday saw a medium where they saw nothing but distance: Faraday sought the seat of the phenomena in real actions going on in the medium, they were satisfied that they had found it in a power of action at a distance impressed on the electric fluids.'

4.4.1   Field Theory

On the wall of his study in Berlin in the 1920s, Albert Einstein had three portraits: Isaac Newton, James Clerk Maxwell, and Michael Faraday. Einstein believed that the greatest change in the intellectual framework of physics since Newton occurred through Faraday's experimental and Maxwell's theoretical work.

Newton showed that terrestrial mechanics and celestial mechanics are synonymous. The paths of celestial bodies such as planets, comets, and space vehicles, as well as the precise times of sunrise and sunset, may be computed via Newton's laws. So may the ebb and flow of the tides on all the shores of the oceans of the Earth. But, Newton's laws and Newton's physics do not help us one iota in accounting for the transmission and reception of radio waves, for the operation of the fax machine, for wireless telegraphy, television, or digital video display (DVD), or for the functioning of cellular phones—nor do they explain how, in the modern electronic age, we may, if we wish, be suffused with the magic of Schubertian

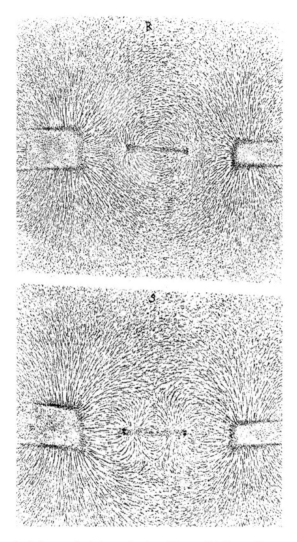

Figure 4.8 *Faraday's famous depiction, using iron filings, of his lines of force and his demonstration that a magnetic field exists outside the perimeter of a magnet.*

(By kind permission of the RI.)

music or the lyricism of the Kreutzer sonata. All of these may be traced back, step by step, to the discoveries of Michael Faraday.

4.5   The Faraday Effect

In 1845, Faraday made his historic discovery that the plane of polarization of a beam of light, on passing through a slab of glass, could be rotated by the application

| (a) **Sir Isaac Newton** (1642-1727) | (b) **James Clerk Maxwell** (1831-1879) | (c) **Michael Faraday** (1791-1867) |

Figure 4.9 *(a) Sir Isaac Newton (Courtesy Trinity College, Cambridge); (b) James Clerk Maxwell (Courtesy Cavendiah Laboratory, Cambridge); and (c) Michael Faraday (Courtesy RI).*

of a magnetic field. This experiment proved that every beam of light has a minute magnetic—and also a minute electrical—component. This is the so-called Faraday Effect in magneto-optics. (With its aid, one may nowadays construct ultra-fast switches in electronic circuitry involving light beams.)

A few weeks after he made this discovery, he despatched to the Royal Society a paper entitled 'On the Magnetization of Light and the Illumination of Magnetic Lines of Force', which begins with a sentence of Chekhovian timelessness: *'I have long held an opinion, almost amounting to conviction, in common I believe with many other lovers of natural knowledge, that the various forms under which the forces of matter are made manifest have one common origin; or, in other words, are so directly related and mutually dependent, that they are convertible, as it were, one into another, and possess equivalents of power in their action....'*

Here is a reflection of his religious conviction. He read the book of nature written by the finger of God, alongside the direct word of God, the Bible.[8,11]

4.6   Three Women with Whom He Interacted

The first of these three Jane Haldimand Marcet, through her book *'Conversations in Chemistry'*, described in Chapter 3, was influential in teaching him the rudiments of Chemistry. Her successful book, based in part on what she had learnt from Davy's lectures at the RI, was read by Faraday when he worked as a bookbinder's apprentice.

The second, Mary Somerville, an excellent mathematician who translated Laplace's work into English, and after whom Somerville College, Oxford, is named, sent letters of admiration to Faraday. In a letter to her composed 1 March 1834, he writes:

Figure 4.10 *Mary Somerville (left), mathematician and popularizer of science, and (right) Ada, Countess of Lovelace, pioneer of computer science, and Lord Byron's daughter.*
(Wikipedia open access)

> 'Dear Madam
>
> 'I cannot refuse myself the pleasure any longer of thanking you for your kindness in sending me a copy of your work. I did intend to read it through first, but I cannot proceed so fast as I wish because of constant occupation.
>
> 'I cannot resist saying too what pleasure I feel in your approbation of my late "Experimental Researches". The approval of one judge is to me more stimulating than the applause of thousands that cannot understand the subject.'

Ada, the Countess of Lovelace, the third of the three Women, was Lord Byron's daughter, and she was an exceptionally good mathematician who took an early interest in the forerunners of modern-day computers.[18] The Countess, so it seems from extant letters she sent to Faraday, was infatuated with him. At one time she pleaded with him to let her collaborate in his laboratories in repeating all his key experiments, a request that could not have been easy for him to deflect. She also told Faraday that she wished to be his 'bride in science'.

4.7 Faraday's Visits to Wales

It is often believed that because of Faraday's prodigious output of original papers, as well as the numerous letters he wrote and received, and because also of his appearance in Government circles, in addition to his role as scientific advisor to the National Gallery, to Trinity House, and the weekly lectures that he gave in Woolwich Arsenal, that he had very little time for recreation.

Yet, in 1819, a year in which he published twelve original papers (he had published twenty-nine such papers the year before), he found time to make his first visit, a walking holiday, to Wales. He left London by coach on 18 July, and then proceeded to the Bristol area, where he crossed the River Severn by boat into Wales. Figure 4.11 shows the route that he then took before he returned by coach via Worcester on 3 August. A daily walk for Faraday often exceeded forty miles.

A former electrical engineer, of Welsh birth, Dafydd Tomos,[17] consulted the copious notes that Faraday made during his walking holiday in Wales that are lodged in the Institute of Electrical Engineers. Not only did Faraday wish to see the beauty of the countryside in Wales—for example, the Dan-yr-Ogof caves and waterfalls nearby in the Neath Valley, the lofty mountain of Cader Idris in mid-Wales, and the Menai Straits separating Anglesey from Caernarvonshire in North Wales—he wanted also to visit the industrial centres located mostly in South Wales and one in Anglesey. Tomos gives detailed accounts of his journey, and even more details made by Faraday, and of Faraday's reaction to all the things he saw. He noted that in the Dowlais ironworks close to Merthyr Tydfil, there were more than fifty miles of tram roads. (This rivals in magnitude the internal railway system now present in Leverkusen, Germany, of the Bayer Company). He also records that in the village of Ystalyfera, in the Swansea Valley, three-thousand men were then employed in the ironworks there.

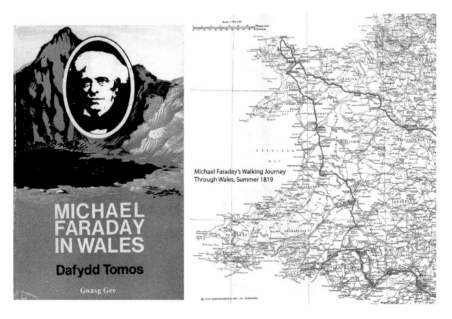

Figure 4.11 *The cover of the book by Dafydd Tomos, showing the lake on the mountain called Cader Idris in mid-Wales. On the right, marked in red, is the route of Faraday's prolonged walking journey.*

(Copyright Dafydd Tomos)

Table 4.2 *Papers Published by Faraday in 1819*

'Carburetted Hydrogen'	Analysis of a stone used in the setting of fine cutlery
'Manganese'	An analysis of Wootz of Indian Steel
'Nitrous Oxide'	On Sirium or Vestium
'Separation of Manganese From Iron'	Palm wine
'Gunpowder Inflamed Without a Spark'	Pyrometrical gauge for a wind furnace
'Some Experimental Observations on the Passage of Gases Through Tubes'	Strength of Aetna wines
'On the Forms of Matters'	

But it was in Swansea and its environs (Hafod and Penllergaer) that he entered into detailed scientific conversations with the owners of the copper-smelting works, which generated an unpleasant sulphurous fog on all the villages in the locality. He stayed in Singleton Abbey, in Swansea (now the site of Swansea University), where he wrote a long letter to his wife, Sarah, telling her how much he missed being away from her.

He revelled in the physical beauty of the scenery that he walked through, such as Ynysygerwn lake in the Vale of Neath, as well as the numerous beauty spots of mid and North Wales.

His visit to Parys Mountain, near Amlwch, which lies at the northerly tip of Anglesey, was of special interest to him, for he explored the copper mines there that had operated for many decades. He went underground, and slid through narrow tunnels of less than two feet diameter. And he comments on how hard the employees worked there. (It is known that some of the miners after working ten hours a day for six days occasionally returned home at the end of the week in debt to their employer and mine owners, as the latter sold the implements that the workers had to use.)

Faraday made two other visits to Wales, one in 1822, the other in 1848. The purpose of his 1822 visit involved the Vivian and Dillwyn families of Swansea, who owned the copper and other (polluting) works in the town's vicinity. Close to the suburb of Landore (the anglicized name for the Welsh Glandwr, 'on the water's edge') were the Hafod and Middle Bank works. Figure 4.12 conveys just how polluting these works were. The owners consulted Faraday to try to meliorate the environmental damage done by their works.

The second visit in 1848 was to attend the meeting of the British Association for the Advancement of Science held in Swansea that year. It was masterminded by an eminent native of Swansea, William Robert Grove (see Chapter 6). Grove was a close friend of Faraday's. He was also Vice-President of the RI. Many of the delegates arrived in Swansea by boat across the Bristol Channel from Ilfracombe, Devon, as the railway lines to Swansea had not been constructed at that time.

Hafod & Middle Bank Works around 1860.

"It came to pass in the days of yore,
the devil chanced upon Landore
quoths he — by all this fume and stink
I can't be far from home. I think!"

Figure 4.12 *Illustration of how polluting the stacks of the copper works near Swansea were in 1860. The author of the doggerel below the image is unknown.*

(With thanks to Professor Keith Smith)

Looking back at the notes composed by Faraday on his three visits to Wales, one is struck by how detailed and informative they are. Some passages are aridly factual; others constitute an invaluable record of industrial activity, and are much coveted by industrial archaeologists.

4.8   Concluding Remarks

There are many more aspects of Faraday's work that need not be elaborated on here. Some were discussed in Chapter 1, such as his introduction of the RI Christmas Lectures, and the Friday Evening Discourses. Other aspects of his work are given in my book.[7] A measure of Faraday's devotion to the Discourses is reflected in Tables 4.3 and 4.4, where a small fraction of those that he delivered and arranged are enumerated.

Reverting to the parallels that may be made between the work of Faraday and Franklin, we are struck by the supreme technical virtuosity, intuition, and genius for experiment that they each possessed. In compliance with Franklin's memorable dictum concerning the necessity not (if one loves life) to squander time,

Table 4.3 *A Selection of Friday Evening Discourses Given by Michael Faraday 1836–61*

Date	Topic	Date	Topic
January 1836	'Silicified Plants and Fossils'	January 1846	'Magnetism and Light'
February 1836	'The Magnetism of Metals as a General Character'	June 1848	'Conversion of Diamond Into Coke'
April 1836	'Plumbago and the Manufacture of Pencils from it'	April 1851	'On Atmospheric Magnetism'
June 1837	'Early Arts: The Bow and Arrow'	June 1852	'On the Physical Lines of Magnetic Force'
		March 1860	'On Lighthouse Illumination—the Electric Light'
February 1838	'The Atmosphere of This and Other Planets'		
April 1842	'Conduction of Electricity in Lightning Rods'	February 1861	'On Platinum'

Table 4.4 *A Selection of Friday Evening Discourse Arranged by Michael Faraday 1832–61*

Date	Lecturer	Topic
March 1832	G. Foggo	'The Causes of the Excellence of Grecian Art'
April 1832	Marshall Hall	'The Laws Which Govern the Mutual Relation of Respiration and Irritability'
April 1834	John Davidson	'The Pyramids of Egypt'
May 1834	Dionysus Lardner	'Babbage's Calculating Machinery'
May 1834	John Dalton	'On the Atomic Theory of Vapours'
May 1836	T. J. Pettigrew	'The Opening of an Egyptian Mummy'
May 1837	Gideon Mantell	'The Iguanadon and Other Fossil Remains Discovered in the Strata of Filgate Forest'
May 1837	M. De La Rue	'The History and Manufacture of Playing Cards
May 1838	John Landseer	'The Astronomy of the Book of Job'
March 1841	J. J. Cooper	'Elkington's New Method of Plating and Gilding'
February 1843	Sir William Robert Grove	'The Gaseous Voltaic Pile'
March 1843	Owen Jones	'Moorish Architecture as Illustrated by the Alhambra'
April 1847	Sir Charles Lyell	'The Age of Volcanoes of Auvergne as Determined by the Remains of Successive Groups of Land-Quadrupeds'
May 1847	Tom Taylor	'The Saxon Epic—Beowulf'
February 1848	Sir Charles Lyell	'The Fossil Footmarks of a Reptile in the Coal Formations of the Alleghany Mountains'
May 1851	Sir Henry C. Rawlinson	'A Few Words on Babylon and Nineveh'
May 1855	Sir James P. Lacaita	'On Dante and the "Divine Commedia'
June 1855	Sir Henry C. Rawlinson	'On the Results of the Excavations in Assyria and Babylonia'
April 1856	Sir Charles W. Siemens	'On a Regenerative Steam Engine'
January 1857	Reverend Fredrick C. Maurice	'Milton Considered as a Schoolmaster'
February 1860	Thomas H. Huxley	'On Species and Races and their Origin'
May 1860	Lord Kelvin	'On Atmospheric Electricity'
April 1861	Hermann von Helmholtz	'On the Application of the Law of the Conservation of Force to Organic Matter'
April 1861	John Ruskin	'On Tree Twigs'
May 1861	James Clerk Maxwell	'On the Theory of Three Primary Colours'

Faraday pursued all that he did with indefatigable zeal, bordering at time on obsession. His perennial dictum was *'work, finish, publish'*. [2][19] He, like Franklin, left the world a better informed place.

Also like Franklin, Faraday had an acute sense of public responsibility. He served as adviser to both the National Gallery and the British Museum, and he responded to requests from Parliament to serve as a member of certain critical investigations. In 1836, he was appointed to a post which gave him considerable pleasure for nearly thirty years: Scientific Advisor to Trinity House. This was the sort of work he enjoyed, involving experimental work, both in the laboratory and in the field, on such things as the design of lighting systems for lighthouses, fog-warning systems, and so on. He continued actively in this work after he resigned all his other appointments, not relinquishing it until 1865, when he handed over to his RI successor, John Tyndall. (Interestingly, a subsequent RI Professor, Lord Rayleigh, took over in his days as Scientific Advisor to Trinity House—see Chapter 5.)

4.8.1 An Assessment of Faraday and Davy

The following comparison was drawn to my attention by the late Professor Ronald King. It has been quoted often at the RI, and I first came across it in the late 1980s. I do not know who the author is. But the sentiments expressed in it are those with which most members of the RI, past and present, would agree.

> "Wherever a true comparison between these two Nobles of the (Royal) Institution can be made, it will probably be seen that the genius of Davy has been hid by the perfection of Faraday. Incomparably superior as Faraday was in unselfishness, exactness and perseverance, and in many other respects also, yet certainly in originality and eloquence he was inferior to Davy, and in love of research he was by no means his superior. Davy from his earliest energy to his latest feebleness, loved research, and notwithstanding his marriage, his temper, and his early death, he first gained for the RI that great reputation for original discovery which has been and is the foundation of his success."

Professor Frank James, formerly of the RI, now at University College London, has drawn to my attention a description that an ardent member of the RI, Juliet Pollock, made about Faraday's lectures in the late 1850s.[16]

'It was an irresistible eloquence, which compelled attention. There was a flame in his eyes which no painter could copy, which no poet could describe. His enthusiasm seemed to carry him to the point of ecstasy when he expatiated on the beauty of nature and when he lifted the veil from its deep mysteries. His light, lithe body seemed to quiver with its eager life. His audience took fire with him, and every face was flushed.'

J. Pollock, *St Paul's Magazine*, **1870**, 6, 293

4.8.2   What was the Essence of Faraday's Genius?

Whenever I give my lecture the 'genius of Faraday',[20] I am invariably asked at the end of my talks to elaborate on the wellsprings of his uniqueness. How could he have accomplished so much? To say he was an experimenter of genius is to risk exhibiting him as little more than a skilled manipulator. That he certainly was, but one for whom skill was the servant of imagination. His true genius lay in his ability to notice some oddity, to devise some experiments, to test its significance, and, with astonishing economy of effort, to discover how, if at all, his picture of the physical world must be modified. And though he never mastered anything beyond the elements of arithmetic, his mode of working was exemplary. As a researcher, discoverer, and expositor he excelled because:

- he possessed unquenchable curiosity
- he had a passion for clarity—in the conception, execution of his experiments, and in describing them
- to all questions that he posed, he believed there were answers
- his choice of problems was astute—he raised important, fundamental questions
- his tactics, strategy, and economy of effort in answering questions were impeccable
- he devised the best possible equipment, instruments, and materials—his coulometers were more sensitive, his electromagnet more powerful, his glass specimens were of superior quality and heavier than those of his contemporaries
- he demonstrated his discoveries (and those of others) to lay audiences and children, in memorable, dramatic ways
- he popularized science to children and adults in graphic, gripping, and eloquent terms

In summary, we see that Faraday combined in a singular fashion, supreme intellectual power, profound intuition, exceptional technical virtuosity, and an almost timeless (Chekhovian) style of describing his work. To cap it all, he was morally incorruptible.

Figure 4.13 *Photograph of Faraday's laboratory (as reconstructed by Professor Ronald King, 1973), with a magnified image, top right, of Faraday's electromagnet.*

(Courtesy Shell Education Brief)

4.8.3 Faraday's Skill in Coining Words

On the £20 note that was issued by the Bank of England to mark the bicentenary of Faraday's birth in 1991, there are several scientific items on display. Fringing the monetary note are lines of force passing from a positive to a negative centre. The equipment that he used for the discovery of electromagnetic induction is shown on the lecturer's bench in the RI. Also clearly shown are the terms in electrochemistry now universally used (see Figure 4.14).

Faraday coined some of these words in association with the polymathic Master of Trinity and Professor of Mineralogy Dr William Whewell, FRS, and also a Greek-Hebrew scholar, Dr Nicalles, FRS.

APPENDIX

Excerpts from the Sermon by the Archbishop of York, Preached at Westminster Abbey at Faraday's Bicentenary Celebrations in September 1991

'I was tempted to use a text from the Book of Genesis: "There were giants in the earth of those days." For today we celebrate a scientific giant, a man whose genius has shaped the modern world. The father of electricity, one of the pioneers of analytical chemistry, organic chemistry, and physical chemistry. A daring thinker who invented the concept of a field of force. A brilliant experimentalist. An example to all scientists of meticulous concern for fact. "I could trust a fact," he said. A great communicator, a teacher of the

Figure 4.14 *A selection of the words coined by Faraday (see text). These are listed on the £20 note printed in his honour.*

young, whose lectures at the Royal Institution were a byword for careful preparation, eloquence, and intelligibility. The founder of a great tradition, which continues to this day.

'And beyond all this, his contemporaries knew him as a great and good man, a man not only unsurpassed in his success as a scientist, but also a shining example of the kind of man a scientist should be.

'A giant. Yet, he would have hated the description. He would have hated its misuse of Scripture, because the literal interpretation of Scripture meant a great deal to him. He would have hated its implications of worldly honour. Though showered with honours from every part of the world, though he rose from obscurity to become respected and sought after by the highest in the land, he remained plain Michael Faraday, more at home in his strange religious sect than in the modern world he helped create. I suspect, too, he would have hated this celebration, so alien to everything his own tightly controlled and narrowly circumscribed religious faith had taught him.

'So let me turn to a text which would have been more appropriate, indeed one which he himself used on a number of occasions.

'Romans 1: "The invisible things of him from the creation of the world are clearly seen, being understood by the things that are made, even his eternal power and Godhead."

'There we find the religious roots of his scientific commitment. A deep belief in the order and intelligibility of the world, a belief that "the invisible things" can indeed be "clearly seen" through "the things that are made". And for Faraday this belief that the world was created as an ordered whole provided the stimulus to go on seeking connections between things which didn't at that time seem to have any clear connections: between electricity and magnetism, between electricity and chemistry; and ultimately as the most daring vision of all, between electromagnetism and gravity.'

REFERENCES

1. M. Faraday, *Phil. Trans. R. Soc.*, **1850**, 1–122.
2. S. P. Thompson, *'Michael Faraday: His Life and Work'* Cassell & Co. Ltd., London, **1901**.
3. R. Appleyard, *'A Tribute to Michael Faraday'*, Constable & Co., **1931**.
4. J. Kendall, *'Michael Faraday: Man of Simplicity'*, Faber & Faber, London, **1954**.
5. H. Bence Jones, *'The Life and Letters of Faraday'*, Longmans, Green & Co., **1870**; F. A. J. L. James, *'The Correspondence of Michael Faraday'*, I–VI, **2018**.
6. L. Pearce Williams, *'Michael Faraday (1791–1867)'*, Leonard Parsons, London, **1924**.
7. J. M. Thomas, *'Michael Faraday and the Royal Institution: The Genius of Man and Place'*, IOP and Adam Helger, Bristol & London (now published by Taylor and Francis), **1991**.
8. C. A. Russell, *'Michael Faraday: Physics and Faith'*, Oxford University Press, Oxford, **2000**.
9. Frank A. J. L. James, *Physics World*, **1991**, September, 4.
10. Often referred to as the birth of electrical engineering.
11. R. King, *'Michael Faraday of the Royal Institution'*, Royal Institution, London, **1973**.
12. J. M. Thomas, 'The Extraordinary Impact of Michael Faraday on Chemistry and Related Subjects', *Chem. Commun.*, **2017**, *53*, 9179.
13. R. Feynman, *'Thoughts of a Citizen Scientist'*, Basic Books, New York, **2005**, pp. 14–15.
14. See ref [7], p. 51.
15. A good description of the Faraday Cage and its use in modern life is available on Google.
16. Professor Frank A. J. L James, who has read all the extant letters written and received by Faraday, as well as numerous articles by him, claims that Faraday regarded the Bible as the word of God, and that all natural phenomena as the product of the hand of God, and that, in all his studies, he sought to uncover the nature of God's creation. (See also the Appendix to this Chapter.)
17. Dafydd Tomos, *'Michael Faraday in Wales'*, Gwasg Gee, Wrexham, **1980**.
18. When, some fifty years ago, the US military devised a new computer language they named it ADA in her honour.
19. This was the advice that Faraday gave to an enquiry that William Crookes made in his early twenties to the great man (see ref [2], p. 267).
20. Which I have done some three-hundred times.

5

The Incredible Lord Rayleigh

5.1   A Cruise up the Nile

The Third Baron Lord Rayleigh, born John William Strutt, last of the great British classical physicists, made contributions to every single branch of the physical sciences known in his day. To gauge quite how remarkably able he was, consider the fact that, aged twenty-six, his health broke down, so he and his wife went on a prolonged convalescent cruise up the River Nile from Cairo. While on that rehabilitating journey he wrote *'A Treatise on Sound'*, a textbook still used by university students. The first volume was published as early as 1877, when Rayleigh had no access to any library, and when he solved all the necessary mathematical equations during the course of his cruise. The second volume appeared a few years after his return to England.

His penetrating insight and prodigious capacity for detailed work enabled him to solve problems previously perceived by his progenitors and contemporaries as intractable, as well as to suggest new lines of research that engendered the blossoming of much of twentieth- and twenty-first-century science and technology.

He did not become the Professor of Natural Philosophy at the Royal Institution (RI) until 1887, having earlier been the successor to James Clerk Maxwell as Cavendish Professor of Physics at Cambridge University from 1879–84.

Here, we shall concentrate predominately on the work, which he did largely at the RI, but also at his own laboratory in his baronial home in Terling, Essex (Figures 5.1 and 5.2).

Only a fraction of Lord Rayleigh's accomplishments are discussed here. There have been several full accounts of all his work.[1–4] We shall deal briefly with some of his afternoon lectures and Friday Evening Discourses at the RI, but at some length with the work done there that earned him his Nobel Prize for Physics in 1904 for the discovery of Argon.

5.2   Brief Outline of his Career

No name occurs more frequently in relation to phenomena, principles, and effects with which any student of classical physics must become acquainted than

Albemarle Street: Portraits, Personalities, and Presentations at the Royal Institution. John Meurig Thomas,
Oxford University Press. © Sir John Meurig Thomas 2021.
DOI: 10.1093/oso/9780192898005.003.0005

Figure 5.1 *(a) Photograph of the twenty-six-year-old Lord Rayleigh at the time of his Nile cruise. (b) Portrait of the Third Baron Rayleigh, OM, President of the Royal Society 1905–08).*

(a) and (b) by kind permission of Lord and Lady Rayleigh

Figure 5.2 *Terling Place, home of the Strutt family. The laboratories created by Lord Rayleigh are on two floors in the near-most wing of the house.*

(By kind permission of Professor E. A. Davis)

that of Rayleigh: Rayleigh Scattering (of electromagnetic waves and the explanation of the blue sky and red sunset),[5] Rayleigh Waves, Rayleigh Criterion (governing resolving power in microscopes and telescopes), Rayleigh Number (in convection), Rayleigh Disc (for measuring the absolute intensity of sound), Rayleigh Fading and Rayleigh Distance (terms used in the propagation of electromagnetic waves), Rayleigh Damping, and the Rayleigh—Jeans Law (for black-body radiation); these are by no means an exhaustive list of the impact of his work, yet they reflect but a fraction of the fields and phenomena in which his interests ranged with the most fruitful of results. For example, in 1885, Rayleigh published an article on the propagation of surface acoustic waves. This was a seminal paper that, for modern-day seismologists and earth scientists, is the basis for detecting and pinpointing the location of distant earthquakes, and for electronic engineers forms the basis for practical delay lines in circuits used in radar and television.

Born on 12 November 1842, the young Strutt entered Trinity College, Cambridge, in October 1861, and was soon following the rigorous courses in mathematics given by E. J. Routh (of Peterhouse), a remarkably successful tutor, who also taught J. J. Thomson and A. N. Whitehead. Strutt was 'senior wrangler'—the term used in Cambridge University to describe the top person graduating in mathematics—in January 1865. Sir James Jeans, the eminent astronomer and cosmologist and Visiting Professor at the RI, writing some sixty years later, said that *'there still lingers in Cambridge a tradition as to the lucidity and literary finish of his answers in the examination.'* The fine sense of literary style which Strutt displayed, even under pressure in the examinations, never deserted him. Every paper he wrote—and there were four-hundred-and-forty-five over a fifty-year period, some dealing with the most abstruse subjects—is a model of clarity, and conveys the impression of having been written with effortless ease.

5.3   Professor at Cambridge

In 1879, James Clerk Maxwell, the first occupant of the Cavendish Chair of Experimental Physics at Cambridge, died of intestinal cancer. Rayleigh agreed to serve as the second Cavendish Professor for the period 1879–84. He took his university duties very seriously, both with respect to the instruction of students and to the carrying out of a vigorous research programme that set about redetermining the values of electrical standards (the ohm, volt, and amp). A classical series of papers resulted from this ambitious project. But, after a five-year tenure, he returned to his laboratory at Terling Place.

Before we allude to his subsequent work and achievements at Terling and the RI, it is instructive to cite the topics of the hundreds of publications—largely single-authored—that Rayleigh published over his career, and these can be seen below in Figure 5.3.

```
┌─────────────────────────────────────────────────────────┐
│                                                         │
│        SCIENTIFIC PAPERS CLASSIFIED ACCORDING TO        │
│                      SUBJECT                             │
│                                                         │
│  MATHEMATICS  25                                        │
│      GENERAL MECHANICS  42                              │
│         ELASTIC SOLIDS  21                              │
│            CAPILLARITY  32                              │
│               HYDRODYNAMICS  85                         │
│                  ELECTRICITY AND MAGNETISM  91          │
│                     DYNAMICAL THEORY OF GASES  21       │
│                        PROPERTIES OF GASES  34          │
│                           THERMODYNAMICS  27            │
│                              SOUND  131                 │
│                                 OPTICS  148             │
│                                    MISCELLANEOUS  36    │
│  445 publications in total                              │
│                                                         │
└─────────────────────────────────────────────────────────┘
```

Figure 5.3 *Breakdown of Lord Rayleigh's publications into different topics.*
(By kind permission of Professor E. A. Davis)

Other notable achievements of Rayleigh included being elected Fellow of the Royal Society in 1873; he served as President of the Royal Society from 1905–08. He was also among the original recipients of the Order of Merit (OM) in the 1902 Coronation Honours.

5.4   Rayleigh at the RI

Both as Professor of Natural Philosophy, and in his capacity as Co-Director of the RI, with Sir James Dewar, Rayleigh gave numerous afternoon lecture-demonstrations and Friday Evening Discourses on topics such as the limits of audition, polish, shadows, the flight of birds, the composition of water, fluid motion, radioactive changes in the Earth, and several others beside. He also delivered fifteen Discourses, including one on Thomas Young and his classic in 1895 on argon, which we deal with below. But, one of his favourite lecture demonstrations dealt with the spreading of oil on water, where he conducted a laboratory equivalent of an experiment carried out in London by Benjamin Franklin in 1757.

5.4.1 Pouring Oil on Troubled Waters with Benjamin Franklin

Rayleigh took particular delight in recalling the experiment that Benjamin Franklin carried out in Clapham Common, London, in 1757. The story is a fascinating one, and has been well told by C. H. Giles in the *Society of Chemical Industry* issue

Figure 5.4 *View of the Mount Pond, Clapham Common, looking north-east, the site of Franklin's experiment.*
(From Giles[6])

in 1964.[6] It seems that Franklin, while on duty in London, remembered a passage in Pliny, while he visited Clapham. These are the words Franklin used:

> *'At length being in Clapham where there is, on the common, a large pond* (see Figure 5.4), *which I observed to be one day very rough with the wind, I fetched out a cruet of oil, and dropt a little of it on the water. I saw it spread itself with surprising swiftness upon the surface.*
>
>   *'I then went to the windward side, where (the waves) began to form; and there the oil, though not more than a teaspoonful, produced an instant calm over a space several yards square, which spread amazingly, and extended itself gradually till it reached the lee side, making all that quarter of the pond, perhaps half an acre, as smooth as a looking glass.'*

5.4.2 Franklin's Monolayer

If we assume that the thickness of Franklin's most extended film represents the greatest extent to which an olive-oil film will spread spontaneously on water and still be effective in wave-damping, we are in a position to make a rough comparison between Franklin's result and more recent ones undertaken in the laboratory. Eighteenth-century spoons had a capacity of *ca.* 2–2.5 ml. Franklin's estimate[7] was half-an-acre for the ultimate extent of spread on one teaspoonful of oil:

$$\text{Half-an-acre} \equiv 2{,}420 \text{ yd}^2 \equiv 2{,}420 \times 0.9144 \text{ m}^2$$
$$\equiv 2{,}420 \times 0.9144 \times 10^{20} \text{ Å}^2$$
$$\text{One teaspoonful} \equiv 2.0 \text{ ml} \equiv 2 \times 10^{24} \text{ Å}^3$$

Therefore the thickness of film when one teaspoonful covers half-an-acre is:

$$(2 \times 10^{24}) / (2{,}430 \times 0.9144^2 \times 10^{20}) \text{ Å} = 9.9 \text{ Å}$$

Rayleigh used to carry out a neat illustrative version of this experiment on the lecturer's table in the RI. Doing his experiment carefully, he arrived at the thickness of a triolein[9] film on water as 16.3 Å, which is one of the earliest estimates of the dimensions of a molecule.

Elsewhere in his account, Franklin says:

> *'If a drop of oil is put on a polished marble table, or on a looking glass that lies horizontally: the drop remain in place, spreading very little. But when put on water it spreads instantly many feet around becoming as thin as to produce the prismatic colours....'*

Thomas Young, in 1801, quantified these phenomena when he considered them as sessile drops, contact angles, and spreading of liquids on surfaces. In a Discourse given by Lord Rayleigh on the subject *'Thomas Young'*, given in June 1899, he dealt quantitatively with all the relevant phenomena that Franklin had described in his visit to Clapham Common. It is because a monolayer of oil forms on the surface of the rough water that it becomes quiescent and flat.

5.5   The Discovery of Argon

It was at the RI that Rayleigh largely conducted the work that earned him the Nobel Prize in Physics and brought him his greatest fame. He also worked a good deal on the determination of atomic weights of the elements at Terling Place. Rayleigh had long been intrigued by William Prout's hypothesis that the atomic weights should be integral numbers. If we assign hydrogen to 1, oxygen should be 16, but it was not quite so. Was the discrepancy real? Rayleigh therefore determined the densities of hydrogen and oxygen; and then he moved on to nitrogen. In a letter to *Nature* (published 29 September 1892) he wrote:[10]

> *'I am much puzzled by some results on the density of nitrogen, and I shall be obliged if any of your chemical readers can offer suggestions as to the cause. According to the methods of preparation, I obtain two quite distinct values. The relative difference, amounting to about one part in 1000, is small in itself, but it lies entirely outside the errors of experiment, and can only be attributed to a variation of the character of the gas.'*

His two sources of nitrogen had been ordinary air with the oxygen removed by heated metallic copper, and a 'lighter' nitrogen obtained by decomposition of ammonia. No illuminating response came at that time from the chemical public, but his colleague at the RI, Sir James Dewar, said that some of the atmospheric dinitrogen N_2 was in an allotropic state, such as N_3, just as some oxygen exists as ozone O_3. Rayleigh was sceptical about this, and he continued for two years to prepare nitrogen by several methods. Such chemically produced nitrogen was always lighter in density than atmospheric nitrogen. One of Rayleigh's lecture-demonstrations at the RI was a repeat of some of Henry Cavendish's studies, carried out a century earlier, in which a globe of ordinary air was subjected to electrical sparking so as to consume the oxygen as an oxide of nitrogen (that could be absorbed by potash). In this manner, all of the oxygen was removed, except, as Cavendish had noted, a very small residue. In other words, Rayleigh verified the reclusive Cavendish's results that had lain unnoticed in the literature for more than one-hundred.

In addition to obtaining nitrogen from ammonia, Rayleigh also obtained nitrogen from other chemical sources of the element, as shown below. And he measured the atomic weights as set out in Table 5.1 (the numbers being the weight in grams actually contained under standard conditions in the glass globe employed).

There is a difference of about 11 mg, or about 0.5 per cent. Rayleigh expatiated on these puzzling facts in a lecture given to the Royal Society on 19 April 1894. This prompted a member of the audience, William Ramsay, who had been Professor of Chemistry at University College, London, since 1887, to converse with Rayleigh (Figure 5.5). By the end of May that year, Ramsay had shown that nitrogen gas, when repeatedly exposed to heated magnesium (to form the solid nitride) could be made progressively denser.

Continuing his experiments throughout the summer, Ramsay produced in early August a gas, apparently unaffected by further treatment with magnesium. This result led to an exchange of letters. Sir William Crookes examined the residual gas and spectroscopically reported that it was new and quite distinct from nitrogen. Rayleigh told Ramsay that the residue was neither oxygen nor nitrogen. The two immediately joined forces, and on 31 January 1895 announced to the Royal Society the discovery of a new gaseous element, apparently inert chemically, which they called argon (for 'inert, without work') (see Figure 5.6).

Table 5.1 *Atmospheric Nitrogen and Chemical Nitrogen*

Atmospheric Nitrogen		
By hot copper (1892)		2.3103
By hot iron (1893)		2.3100
By ferrous hydrate (1894)		2.3102
	mean	2.3102
Chemical Nitrogen		
From nitric oxide		2.3001
From nitrous oxide		2.2990
From ammonium nitrite		2.2987
From urea		2.2985
From ammonium nitrite purified differently		2.2987
	mean	2.2990

Prof. William Ramsay.   Rayleigh.
Sept. 1894

Figure 5.5 *Sir William Ramsay and Lord Rayleigh (1894). This picture was taken after the isolation of the first noble gas.*
(By kind permission of Lord and Lady Rayleigh)

4th August, 1894

Dear Lord Rayleigh,
    I have isolated the gas. Its density is 19.075, and it is not absorbed by magnesium . . .

6th August, 1894

Dear Prof Ramsay,
    I believe that I too have isolated the gas, though in miserably small quantities . . .

Figure 5.6 *The opening sentences of the letters in which Ramsay and Rayleigh reported their simultaneous isolation of argon.*
(Royal Society)

Rayleigh and Ramsay's fifty-four-page joint paper gave the density, refractive index, solubility in water, ratio of specific heats (C_p/C_v), and atomic spectrum of the new gas, and they postulated a new zeroth column, for noble gases, in Mendeleev's periodic table. Some scientists argued that so heavy an element could not possibly be a gas. Rayleigh, with aristocratic humour, replied:

'...*the result is, no doubt, very awkward. Indeed, I have seen some indications that the anomalous properties of argon are brought as a kind of accusation against us. But we had the very best intentions in the matter. The facts were too much for us, and all that we can do now is apologize for ourselves and for the gas*'.

5.6   Conclusion

Lord Kelvin, a close friend of Rayleigh's (see Figure 5.8), made the below statement shortly after Rayleigh and Ramsay shared their Nobel Prize for Physics in 1904. He hailed argon as undoubtedly the greatest scientific event of the year. Specifically, he said:

'*If anything could add to the interest which we must all feel in this startling discovery, it is the consideration of the way by which it was found—arduous work—commenced in 1882, has been continued for 12 years by Rayleigh, with unremitting perseverance.*'

Figure 5.7 *Painting of the Third Baron Rayleigh by Sir Philip Burne-Jones.*
(By kind permission of Lord and Lady Rayleigh)

In Rayleigh's home at Terling there hangs a beautiful painting by Sir Philip Burne-Jones of the Third Baron Rayleigh working at his laboratory bench (see Figure 5.7). What most scientists find so remarkable about Rayleigh was that he was superbly good at being both an experimentalist and a theoretician.

According to Sir James Jeans, referring again to Rayleigh:

'...his massive, precise and perfectly balanced mind was utterly removed from that of the erratic genius who typifies the great scientist in the public imagination...The outstanding qualities of his writings were thoroughness and clearness; he made everything seem obvious. Rayleigh died in Essex on 30 June 1919, having been at work on a scientific paper only five days previously...The inscription on his memorial in Westminster Abbey, "An unerring leader in the advancement of natural knowledge", does not overstate the case.'

Figure 5.8 *The Lords Rayleigh (left) and Kelvin (right), two of the foremost scientists of the nineteenth century at Terling Place.*
(By kind permission of Lord and Lady Rayleigh)

APPENDIX 1

A Wager Involving the Age of the Earth

When Ernest Rutherford gave a Discourse at the RI in early 1904, while still a young professor in McGill University, Montreal, he forwarded cogent (radioactive) evidence to suggest that the age of the Earth was at least eight-hundred million years (its actual age is four-and-a-half billion years). Lord Kelvin and Lord Rayleigh were both in the audience. After the lecture, Rayleigh told Kelvin that he was far more inclined to believe Rutherford's estimate than Kelvin's earlier one of about forty million years. Rayleigh told Kelvin that he would put on dinner in Terling Place to bring Kelvin and Rutherford together. Rayleigh had laid a bet of £5 that Kelvin would come round to Rutherford's view. Figure 5.9 is a copy of the Terling Place visitors' book, showing who came as guests later in 1904.

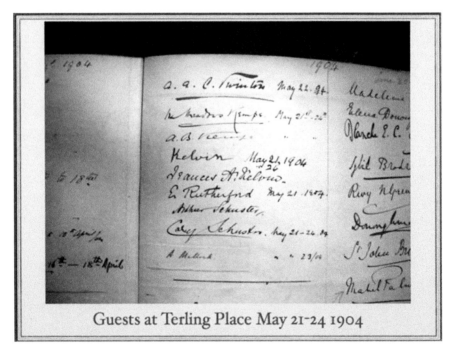

Guests at Terling Place May 21-24 1904

Figure 5.9 *Extract from the visitors' book at Terling Place, May 1904, confirming the meeting between Lord Kelvin, Ernest Rutherford, and Professor Schuster.*
(By kind permission of Lord and Lady Rayleigh)

APPENDIX 2

Rayleigh and Ramsay

William Ramsay, to my knowledge, never gave a Discourse at the RI. He became very friendly with Rayleigh, but, from what I heard at the RI, there was no love lost between Ramsay and his fellow Scottish scientist James Dewar.

Shortly after Rayleigh and Ramsay made their historic announcement about argon in 1895, Sir Henry Miers, of the British Museum, suggested that argon might be identical with an inert gas, supposedly nitrogen, that had been obtained by heating certain uranium-containing minerals. Ramsay prepared this gas and found that it was not argon, but yet another new gas. Crookes identified it as the element helium, which had been identified spectroscopically as being on the Sun's surface in 1868.

The atomic weights of the new gases confirmed Ramsay's early suspicion (1892) that there is room for several gaseous elements at the end of the first column of Mendeleev periodic table. In due course, Ramsay and Morris Travers, on fractionally distilling liquefied air, isolated neon, krypton, and xenon. Finally, with Whytlaw-Gray, Ramsay identified, using a minute sample, the last member of the family: this was radon, a product of radioactive decay. So, between them, Rayleigh and Ramsay were instrumental in significantly expanding the periodic table (Figure 5.10).

APPENDIX 3

Message from the Current Lord Rayleigh

'I met Sir John Meurig Thomas when he kindly invited me and my uncle, Guy Strutt, to the Royal Institution as his guests. He was director at the time; I haven't got an exact date clearly in my head, but I think it was soon after I succeeded from my uncle to the title of Lord Rayleigh (April 1988). Sir John and his wife, Margaret, were most welcoming and we listened to John Howard give a Discourse on my eminent forebear John William (The 3rd Lord Rayleigh).

'My inheritance included Terling Place where my grandfather and great-grandfather had a laboratory located in the West Wing. I am correctly humble about my scientific ability and how it was applied by my forebears at Terling and my concern was that much of the valuable scientific information might be lost with my Uncle Guy (then about 70 and the last living connection with it as a working laboratory).

'Sir John and I met up in The Athaneum, where we hatched a plan to use Royal Institution resources to begin recording what otherwise might be lost to posterity. Initially, Dr Bryson Gore and Mr Coates, from the RI, came down and recorded my uncle explaining the contents and did scoping surveys of what was on display.

'Sir John then secured the help of Harry Roenburg to undertake a more detailed inventory of each item that was on display or in cupboards, including in the basement

Figure 5.10 *Picture by the London cartoonist Leslie Ward ('Spy', 1851–1922); Ramsay points to the eighth group of the periodic table, which only contains elements that he discovered and isolated. (Public domain)*

where the balance chamber for the storage and weighing of argon took place. Harry was semi-retired from The Clarendon Laboratory; he and his wife, Anna, visited Terling regularly often over-nighting with my Uncle Guy. Each article in the lab and basement was given a unique number, there was a schedule appended to each cupboard of all the items contained within, with unique numbers and then an index car with a description, a drawing, and reference to his works (if possible). Sadly, Harry died when he was about 90 per cent through the cataloguing and his colleague from the Clarendon, Neville Robinson, briefly filled the void.

'I then contacted Sir John, who had moved on to become Master at Peterhouse, and he found me Professor (Ted) Davis to take forward Harry's great work. Ted was co-located with Sir John in the Materials Science and Metallurgy Department at Cambridge and was coordinating editor of the Philosophical Magazine.

'Ted is the de facto curator of the laboratories, organizing visits and filtering requests to pay homage to the site where argon was isolated and weighed. He has produced photographic records of the items catalogued by Harry, and gives talks on my forebears' works and tours of the laboratory to interested parties. He continues to expand the use to which certain items were put and the phenomena which my great grandfather's name is associated.'

Figure 5.11 *Photograph of John Meurig Thomas (left) and Professor E. A. Davies in front of the portrait of Lord Rayleigh at Terling Place, 18 March 2017.*

REFERENCES

1. Robert Bruce Lindsay. '*Lord Rayleigh: The Man and his Work*', Pergamon Press, **1966**.
2. J. H. Howard, 'Research of the Third Baron Rayleigh', *Proc. Roy. Inst.*, **1988**, *60*, 73.
3. M. Longair, '*Maxwell's Enduring Legacy*', Oxford University Press, Oxford, **2016**, p. 69.
4. E. A. Davis, 'Lord Rayleigh: His Work and Laboratories', in P. M. Schauster, ed., '*Proc. 1st European Symposium in the History of Physics and Chemical Nitrogen*', Living Edition Science, Austria, **2008**, p. 55.
5. Rayleigh showed that light is scattered by molecules and small particles in the air with an intensity that is proportional to the inverse of the fourth power of the wavelength of light. This accounts for the red of the sunset and the blue of the daylight sky.
6. C. H. Giles, *J. Soc. Chem. Industry*, **1964**, 8 November.
7. B. Franklin, *Phil. Trans. R. Soc.*, **1774**, *64*, 455.
8. Rayleigh, *Proc. Roy. Soc.*, **1890**, *47*, 364.
9. Triolein is the main constituent of olive oil.
10. Rayleigh, *Nature*, **1892**, *46*, 572.

6

The Fuel Cell: William Robert Grove's Discourse in 1843 and Francis Bacon's in 1960

6.1 Introduction: The Threat of Climate Change

In April 2020, a plea was made by David Attenborough on British television when he pointed out the dire consequences that would arise if the problem of climate change were not solved.

Arguably, one of the most important challenges facing the human race is the need to eliminate enormous production of carbon dioxide CO_2, which is responsible for rising global temperatures that cause the melting of polar ice and rises in ocean levels. Future increases in world population will exacerbate this problem because of the (population-related) increasing use of internal combustion engines (ICE) and the extensive burning of fossil fuels. A fundamentally different means of generating energy for transport, lighting, and heating is required. Thanks to the Discourse at the Royal Institution (RI) in 1843 by the lawyer and amateur scientist William Robert Grove, and an extension of it by Francis Bacon in 1960, a solution in principle already exists, and it could lead—provided the practical problems involved are overcome—to a universal hydrogen economy, as explained below.

Grove discovered that his so-called, curious Gaseous Voltaic Pile allowed hydrogen and oxygen at the surface of platinum in a simple electrochemical cell to combine silently and to lead directly to the production of electricity with water as the only by-product. No CO_2 is produced. He shared this discovery with Michael Faraday in 1842 (Figure 6.1.)

My highly summarized argument above focuses on the urgent necessity for the human race to reduce greatly the production of CO_2, and to turn to a viable alternative method of decarbonizing the world's present method of generating energy.

Albemarle Street: Portraits, Personalities, and Presentations at the Royal Institution. John Meurig Thomas, Oxford University Press. © Sir John Meurig Thomas 2021.
DOI: 10.1093/oso/9780192898005.003.0006

Figure 6.1 *Letter sent from the London Institution by W. R. Grove to Michael Faraday in October 1842. The transcript is as follows:* 'My dear Sir, I have just completed a curious voltaic pile which I think you would like to see, it is composed of alternate tubes of oxygen & hydrogen through each of which passes platina foil so as to dip into separate vessels of water acidulated with sulphuric acid the liquid just touching the extremities of the foil as in the rough figure below.'[1] *(Royal Institution (RI))*

In this chapter, I shall first summarize the essence of the fuel cell, which was first elaborated by its inventor, William Robert Grove, at the RI in his 1843 Discourse, after he had written earlier (October 1842) to Michael Faraday.

We shall then describe briefly two subsequent Discourses at the RI in which the viability of the fuel cell was examined by Francis Bacon in 1960[2] under the title *'Fuel Cells: Will They Soon Become a Major Source of Electrical Energy'*, and a subsequent one by K. R. Williams in February 1991 on *'Electric Vehicles–An Answer to the Environmental Challenge'*.[3]

Bacon was responsible for the fuel cell used in the Apollo 11 mission to the Moon in 1969 (see Figure 6.2). It provided energy for the space capsule and also generated drinking water for the crew, as its electricity (see below) came from the electrochemical combustion of H_2 and O_2 to yield water.[3] K. R. Williams was

Figure 6.2 *Man's first step on the Moon; photograph dedicated by the Apollo astronauts to Francis Bacon.*

(By kind permission of Francis Bacon)

once Head of Shell Thornton's Surface Reactions Division that created the world's first liquid-fuelled, complete fuel-cell power source in 1964.[3]

In his State of the Union address of 2003, President G. W. Bush made the following statement: *'With a new national commitment, our scientists and engineers will overcome obstacles to taking these (hydrogen cell cars) from laboratory to showroom, so that the first car driven by a child born today will be powered by hydrogen and pollution free.'*

For numerous reasons, practical and political, this prediction never came to fruition in the US. And in the intervening years, more effort was expended there and elsewhere in developing electric cars that have conventional batteries, rather than fuel cells, the former being cheaper and simpler to construct. But their mileage range is far smaller than those of fuel cell vehicles (FCV).

Key issues associated with the rapid replacement of ICEs by FCVs include finding cheaper, more active catalysts than platinum, adequate supplies of cheaper hydrogen, and safe means of incorporating it and storing it in a vehicle. These issues now demand renewed and sustained attention. A large number of H_2 refuelling centres is also required.

6.2 William Robert Grove at the London Institution

From the London Institution (an establishment formed in 1819, but which ceased to exist more than a century ago),[1] where he had been earlier appointed Professor, Grove wrote the letter shown in Figure 6.1. In February, 1843, Grove gave his Discourse entitled *'Gaseous Voltaic Pile'*, in which a summary of his extensive article in the *Philos. Trans. of the Royal Society*[4] was presented. Fundamentally, the essence of Grove's fuel cell is the electrochemical combination of H_2 and O_2 to yield electricity directly using platinum electrodes, as catalyst. In effect, Grove had discovered a way of taming this violent reaction when it is carried out by bringing the two gases together in the presence of platinum under normal circumstances. But there was a further key factor. Grove had generated electricity by circumventing the thermodynamic restrictions that follow from Carnot's theorem. Continuous electricity, after Faraday's discovery of electromagnetic induction, is produced by first generating steam, which drives a turbine that then generates the electricity. So the fuel cell is thermodynamically a more efficient way of generating electricity from a given source of energy; it is not influenced by difference in temperature, as in conventional generation.

An example of the early descriptions of a fuel cell published by Grove is shown in Figure 6.3. Taken from recent work[5] it is a typical set-up for the so-called proton-membrane fuel cell, shown in Figure 6.4.

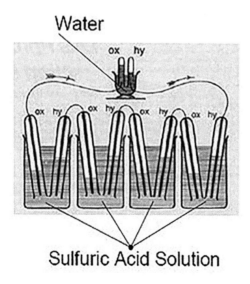

Figure 6.3 *An early version illustrated by Grove of his fuel cell constructed in 1839.*
(Wikipedia open access)

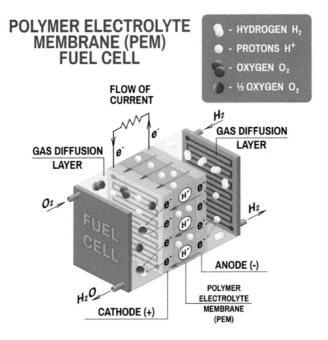

Figure 6.4 *Simplified illustration of a modern hydrogen fuel cell in which the electrolyte is a polymer membrane.*
(Wikipedia open access)

6.3   A Brief History of the Status of the Fuel Cell

There are numerous publications dealing with this topic, both in popular journals such as *Scientific American* and also in several textbooks. In a recent review, the author, with his collaborators Edwards, Dobson, and Owen,[5] has given an account of present, international activity in this important area of clean energy. The US Department of Energy gave a comprehensive record in 2011 of all the various types of fuel cells that can be constructed. Some operate at low, others at high temperature. The nature of the catalyst used to combine the H_2 and O_2 at the respective electrodes can vary widely. But, the biggest problems holding up widespread acceptance of the fuel cell (to decarbonize the present mode of generating energy) are:

- facile and cheap production of clean (green) hydrogen
- safe method of storage of the hydrogen in filling stations and on board vehicles
- extensive network of H_2 refuelling stations

(Green hydrogen is now the term used to describe hydrogen that is produced by renewable sources, such as wind and solar energy.)

One of the most important of the problems holding up widespread acceptance of the fuel cell is the method of producing green hydrogen. It is interesting to note that Bacon, in his 1960 RI Discourse, raised the possibility that nuclear energy should be used to electrolyse water, so as to build up stocks of the H_2 fuel. Very recently Rolls-Royce and other companies have raised this mode of production, in the context of future air travel.

The viability of manufacturing safe means of transport, using zero-emission, hydrogen-fuel-cell motivation is not in dispute. The Chicago Transit Authority, as well as authorities in numerous other cities worldwide including Berlin, have for more than two decades used hydrogen in their fuel-cell-operated city buses (see Figure 6.5).

It is encouraging to note that the 2016 China Technology Road Map calls for a thousand hydrogen refuelling stations (HRSs) to be operative by 2030, with the H_2 production coming from renewable sources. China also plans to have a million FCVs on their roads at that time. In South Korea, it is predicted that there will be more than six-hundred-thousand FCVs serviced by about five-hundred HRSs by 2030. They also will greatly expand the production of heavy-duty FCVs for long-range transport and fork-trucks for sale in the US and Canada. In June 2020, South Korea will build a 50-MW fuel-cell-powered plant that will provide electricity for around eighty-thousand homes. While these developments are encouraging, it must not be forgotten that countries such as China have hundreds of millions of vehicles that currently use fossil fuels and ICEs. This illustrates the magnitude of the political obstacles that have to be overcome.

Figure 6.5 *Cities such as Chicago and Berlin have long used zero-emission, fuel-cell propelled public transport.*

(Wikipedia open access)

The recent report[6] by Hydrogen Europe shows that essentially some of the major private and public organizations in Europe are now fully committed to proceeding towards the hydrogen economy. It is gratifying that now the world's largest companies, such as the oil giants, are actively targeting means of massively increasing clean H_2 production, so as to attack the problem of decarbonizing current processes of generating energy. A notable example is the Californian Fuel Cell Partnership, which aims to see a million zero-emission FCVs on the road by 2030. Also, in the Netherlands, one of the world's first H_2-fuelled power plants is being developed. In Austria, the world's largest green H_2 plant with 6-MW capacity became operational in November 2019. The oil giant Shell, along with Gasunie and Groningen Seaports, plans to build the world's largest offshore wind farm in the Dutch North Sea to produce green hydrogen—and they plan to pursue this despite the present coronavirus crisis.

In the UK and several European countries, strenuous—but not yet adequate—efforts are being made to incorporate the hydrogen economy. Scotland has a Forum for Renewable Energy Development, and it aims to make that country the first in the world to have 100 per cent hydrogen networks. A major achievement in Scotland is the H_2 fuel-cell-powered ferryboat (due to be delivered in 2020) for use around the Orkney Islands, the H_2 being produced locally using wind and tidal sources.

Some key issues that need to be addressed for all countries in order to facilitate the full arrival of the hydrogen economy is to generate H_2 by more efficient electrolysis of water with renewable energy. Geothermal energy in New Zealand will soon produce H_2 there, via electrolysis, and that will be shipped to Japan for everyday use. Likewise, abundant wind energy in countries such as Argentina and Portugal will yield H_2 to be shipped to other countries for fuel-cell use.

The Japanese government has arguably done more than any other to raise the profile of H_2 among global policy makers. It has currently more than a quarter of the world's H_2 filling stations. (It had intended to use 160 unmanned HRSs in time for the 2020 Olympic Games.)

An important CO_2-reducing advance has recently been achieved in Germany and Sweden in the production of steel (which currently liberates *ca.* eight per cent of all global CO_2). In place of using carbon—that has been the practice for centuries—to chemically reduce the iron oxides, it is now feasible to use H_2, thereby comprehensively diminishing the liberation of CO_2. Similar advances are now called for to decrease the enormous amounts of CO_2 liberated in the manufacture of cement.

6.4   William Robert Grove: The Man and Some of his Other Achievements

William Robert was born in Swansea, the son of a local magistrate and Deputy Lieutenant of Glamorgan. He was educated privately, principally by the Reverend

Eli Griffiths. He attended Brasenose College, Oxford, where he studied classics, but it is thought that while at Oxford his mathematical interests were cultivated by Reverend Baden-Powell (Savilian Professor of Geometry and father of Lord Baden-Powell, founder of the Boy Scouts). He graduated in 1832 and in 1835 he was called to the Bar in Lincoln's Inn, London. In that year, he joined the RI and shortly thereafter he was a founder of the Swansea Literary and Philosophical Society—the forerunner of the Royal Institution of South Wales (a body still in existence). Although he had qualified as a barrister, his deep interests at this stage of his career were in natural philosophy, especially in the performance of voltaic cells. While in Swansea, he developed a new battery, which delivered an electro-motive force (emf) of close to 2 volts (Faraday was fond of using it thereafter). He exhibited it at the Académie des Sciences in Paris in 1839, and gave a Discourse on it in the RI in March 1840. He was elected Fellow of the Royal Society (FRS) in 1840, and in 1841 he was one of the founders of the Chemical Society (now the Royal Society of Chemistry).

In 1841, Grove became the first Professor of Experimental Philosophy at the London Institution. In 1840, he did some elegant photography with another FRS, J. D. Gassiot. Grove presciently observed:

> 'It would be vain to attempt specifically to predict what may be the effect of photography in future generations. A process by which the most transient actions are rendered permanent, by which facts write their own annals in a language that can never be obsolete, forcing documents which prove themselves—must inference itself not only, with science, but with history and legislation.'

In a wide-ranging 1842 lecture at the London Institution, entitled '*The Co-Relation of Physical Forces*', a subject which he later developed and expanded with his publication '*The Correlation of Physical Forces*',[7] he dealt with the inter-convertibility of various forms of energy, and gave, in effect, an early account of the first law of thermodynamics: '*energy can neither be created nor destroyed, it can only be converted from one form to another.*'

Grove's book proved a resounding success. It was translated into other European languages and it ran to six editions. Its lucid style is redolent of a deeply cultured and broadly educated man. The first few notes at the end of his book reflect this fact (Figure 6.6).

Many other European scientists, notably Julius von Mayer and Hermann von Helmholtz[8] in Germany, were at that time pursuing, in their own ways, their endeavours concerning the law of the conservation of energy. In the UK and else-where, however, there were several sceptics, and this meant that Grove accumulated critical opposition to his views. In due course, however, he, Mayer, Helmholtz, and others triumphed. It is noteworthy that in later editions of his book, Grove gave a charming account of the situation in which he initially found himself when he proposed his concepts on the correlation of forces—see Figure 6.7.

NOTES AND REFERENCES
TO THE CORRELATION OF PHYSICAL FORCES.

———◦◆◦———

PAGE

5. THE reader who is curious as to the views of the ancients regarding the objects of science, will find clues to them in the second book of ARISTOTLE'S Physics, and in the first three books of the Metaphysics. See also the Timæus of PLATO, and RITTER'S History of Ancient Philosophy, where a sketch of the Philosophy of LEUCIPPUS and DEMOCRITUS will be found.

6. BACON'S Novum Organum, book ii. aph. 5 and 6.

7. HUME'S Enquiry concerning Human Understanding, S. 7, London, 1768.

Figure 6.6 *Excerpts from the endnotes of Grove's book, which illustrates the extent of his learning and general (especially classical) knowledge.*
(Copyright Longman, Green & Co.)

Every one is but a poor judge where he is himself interested, and I therefore write with diffidence ; but it would be affecting an indifference which I do not feel if I did not state that I believe myself to have been the first who introduced this subject as a generalised system of philosophy, and continued to enforce it in my lectures and writings for many years, during which it met with the opposition usual and proper to novel ideas.

Figure 6.7 *A charming and magnanimous paragraph contained in Grove's preface to the fifth edition of his book.*
(Copyright Longman, Green & Co.)

At the London Institution, Grove accomplished much else, as evidenced by the work that he described in his Bakerian Lecture to the Royal Society[9,10] in 1846. He was one of the first, if not the first, to demonstrate that a platinum catalyst served to facilitate the combination of H_2 and O_2 to yield water, as well as facilitating the decomposition (at high temperature) of water into its component elements. (The thermodynamic principle of microscope reversibility, which was not enunciated as such until the 1920s, demands that this be so.) While at the London Institution,

Grove also devised a simple filament lamp, in which a platinum wire, held in a vacuum, lit up when an electric current (delivered by a Grove cell) was passing through it. This was one of the earliest examples of the use of a filament lamp.

Apart from the distinction of being awarded the Royal Society's premier lectureship (the Bakerian), he had the even greater honour of being awarded its Royal Medal. As Vice-President of the Royal Society, he initiated reforms concerning the manner in which the Society conducted its affairs. It was he who was largely responsible for a change in policy as regards the election of new Fellows. That is, a fixed number should be elected in a single year.

6.5 Conclusion

Grove was a remarkable member of the RI: his friendship and common views on the unity of the forces of nature alone endeared him to Faraday, who also admired his experimental ingenuity, especially in regard to the 'intensity' battery that Grove had devised. Grove eventually returned to the law; he became a Queen's Counsel (QC) in 1853, a judge in 1871, and later a Privy Councillor (PC). His hometown, Swansea, is very proud of him. They rejoice that one of their sons could be PC, QC, FRS, and Fellow of the Royal Society of Edinburgh (FRSE), and hold an

Figure 6.8 *Statue and plaque in Swansea, William Robert Grove's hometown.*
(Copyright Professor J. V. Tucker)

honorary doctorate from Cambridge University. First to enunciate the first law of thermodynamics, he was, in addition, instrumental in founding the Royal Institution of South Wales, and he helped bring the **British Association for the Advancement of Science** in 1848 to Swansea. He was also an accomplished astronomer, who did much of his observational work with W. H. Smyth (the father of Piazzi Smyth, mentioned in Chapter 9). Little wonder that a statue (Figure 6.8) of him was established in Swansea a decade ago, principally to record his discovery of the fuel cell, which may still play a vital role in the environmentally responsible creation of energy. A recent comprehensive account of Grove's scientific and other work has been published by Iwan Morus.[11]

REFERENCES

1. Letter from William Robert Grove, Professor at the London Institution, to Michael Faraday at the RI.
2. F. T. Bacon, *Nature,* **1960**, *186*, 589.
3. K. R. Williams, RI Discourse, February 1991.
4. W. R. Grove, *Phil. Mag.,* **1842**, *21*, 417.
5. J. M. Thomas, P. P. Edwards, P. J. Dobson, and G. P. Owen, *J. Energy Chem.,* **2020**, *5*, 405–415.
6. Hydrogen Europe Annual Report, 'Fuel Cell and Hydrogen Joint Undertaking', *in 'Making Hydrogen Energy an Everyday Reality Across Europe'*, **2019**.
7. W. R. Grove, '*The Correlation of Physical Forces*', 6th edn, Longmans, London, **1874**.
8. H. L. F. von Helmholtz (1821–94), German physician, physiologist, and physicist, one of the greatest scientists of the nineteenth century.
9. W. R. Grove, '*On Certain Phenomena of Voltaic Ignition and the Decomposition of Water into its Constituents by Heat*', Bakerian Lecture of the Royal Society, 19 November 1845.
10. He was also Vice-President of the RI. W. R. Grove, *Phil. Trans. R. Soc.,* **1845**, *135*, 351, and J. M. Thomas, *Phil. Mag.,* **2012**, *92*, 3757.
11. Iwan Morus, '*William Robert Grove*', University of Wales Press, Cardiff, **2017**.

7

Molecular Biology and the Crucial Role Played by the Davy–Faraday Research Laboratory in its Birth

7.1 Introduction

It was possibly a gleam in the eye of Sir William Henry Bragg, when, during the course of his Huxley Lecture given at Charing Cross Hospital[1] in 1929, he uttered the following sentence *'I express the hope that (my) team's work on crystal structures of organic substances will, one day, connect up with the medical care of the human body.'* This prophetic hope, it is now fully recognized, has been amply fulfilled.

Molecular biology is currently at the heart of biomedical research, which entails elucidating the chemistry of life. This, in turn, involves determining, in atomic detail, the structure of the macromolecules—enzyme, vitamins, tendons, proteins, muscles, hair, fingernails, and a host of other constituents of living matter, and their role in sustaining the process of life. It is also central to producing pharmaceuticals and medicines that help to conquer disease, and to secure cleaner, safer, and healthier lives.

When William Henry Bragg, as Director of the Davy–Faraday Research Laboratory (DFRL), made the above statement, he and his son, William Lawrence Bragg, who was at that time Head of the Department of Physics in the University of Manchester, had each set up world-renowned centres for the determination of the structures of inorganic and some organic molecular crystals by X-ray diffraction. William Henry Bragg had collected around him in the 1920s a galaxy of bright young physicists (and one chemist)—see Figure 7.1—who were extending further the power of the technique of X-ray diffraction that had been discovered in Munich in 1912 by Max von Laue and his associates.

In expressing the opinion, as he did in his Huxley Lecture, William Henry Bragg, must have been encouraged to do so by the pioneering experiments that two of his acolytes, W. T. Astbury and J. D. Bernal, the former especially, had recorded

Albemarle Street: Portraits, Personalities, and Presentations at the Royal Institution. John Meurig Thomas, Oxford University Press. © Sir John Meurig Thomas 2021.
DOI: 10.1093/oso/9780192898005.003.0007

K. Lonsdale   W. T. Astbury   J. D. Bernal

J. M. Robertson   E. G. Cox

Figure 7.1 *Some of Sir William Henry Bragg's acolytes at the Davy–Faraday Research Laboratory, Royal Institution (RI), London, 1923–40.*
(Courtesy the Royal Society, the RI; copyright Godfrey Argent and Oregon State University)

round about that time, namely that hair, for example, is largely composed of a species of the proteins knows as the keratins (see Figure 7.2).

7.2 Crystals of the Living Body

In a subsequent Friday Evening Discourse at the RI in January 1933 on the topic of *'Crystals of the Living Body'*,[2] William Henry Bragg drew attention to the fact that other proteins (nerves, muscles, and tendons, for example) all possessed arrangements of atoms similar to those of keratin.

He then made the vitally important observation that structure determination of molecules involving X-ray diffraction *'…provides us with a means of examination of*

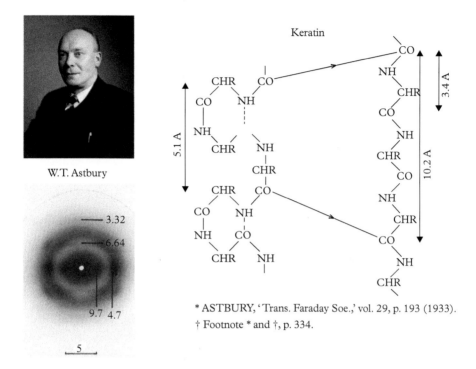

W.T. Astbury

* ASTBURY, 'Trans. Faraday Soe.,' vol. 29, p. 193 (1933).
† Footnote * and †, p. 334.

Figure 7.2 *Structures proposed by W. T. Astbury for both α-keratin (left) and β-keratin (right). (After W. T. Astbury.)*

(Copyright the Royal Society of Chemistry)

structures which are of much greater power than that we possessed previously'. This was an extremely important realization which J. D. Bernal and his Ph.D. student in Cambridge, Dorothy Crowfoot, were later to demonstrate indisputably in their X-ray diffraction study of the digestive enzyme pepsin.[3] In essence, what William Henry Bragg had pointed out was that the methods of the chemist in determining the atomic details of the structure of macromolecules involved in living processes were totally inadequate. In his own words: *'Our chemical methods do not reveal the nature and details of the molecular arrangements. When we employ them for the analysis of a material, we begin by pulling the material to pieces and so destroying the very arrangement of molecules which we should be glad to examine.'* William Henry Bragg had, by this remark, highlighted the then procedure of expert organic chemists in determining the structure of large molecules. Even smallish molecules, like strychnine (the toxic alkaloid), which has a molecular mass of 334, had to be progressively dismantled by the expert organic chemist before its structure could be, in essence, established by re-assembling the smaller component parts, the structures of which could be readily determined. (For more than a century, this is how organic chemists determined the structures of large organic molecules. They revelled in the procedure that was often called *'the process of exhaustive degradation'*.) But, by 'large' we mean, in the context of molecular biology, those of quite modest size. Nowadays, thanks to the pioneering thoughts of William Henry Bragg, and later the architects of structural biology,[3] organic chemists and other scientists dealing with macromolecular entities resort to well-tried methods based on X-ray diffraction to determine their structures.

John Kendrew and Max Perutz, who solved the structures of the giant proteins myoglobin (MW 16,800 D) and haemoglobin (MW 68,000 D), would never have solved these detailed structures in the 1950s if they had relied solely on the exhaustive degradation and re-assembly approaches of the natural product breed of organic chemists.

7.3 The Contributions of Astbury and Bernal

Even after they departed from William Henry Bragg's DFRL (in 1928, to Leeds University in the case of Astbury, to take up the Chair of Biomolecular Materials and to Cambridge University, in 1927, when Bernal was appointed lecturer in crystallography in the Department of Mineralogy and Petrography), these two colleagues carried on their collaboration in the embryonic field of molecular biology. Their work has been fully described elsewhere.[4] Suffice it to note that Astbury expanded greatly his study of proteins, and Bernal pursued the structural studies of numerous types of 'living molecules', including amino acids, vitamins, steroids, sex hormones, and a wide range of proteins. He also started to investigate viruses in his early days at Cambridge.

In due course, Cambridge, Leeds, and Oxford University (to where Dorothy Crowfoot, later Hodgkin, returned after the completion of her doctorate) all became lively centres of research in molecular biology, which also continued with reduced vigour at the DFRL with the onset of World War II in 1939.

In the US also, work began at the California Institute of Technology under Linus Pauling, who was supported by the Rockefeller Foundation. (In fact, the first use of the term 'molecular biology' was by Dr Warren Weaver, who was the Director of Natural Sciences at the Rockefeller Foundation.[5])

In 1938, Bernal, Perutz, and Isidor Fankuchen began[6] studying the structure of haemoglobin, as Bernal had convinced his contemporaries, thanks to his study of pepsin[7] with Crowfoot in 1934, that the riddle of life was hidden in the structure of proteins.

It took nearly twenty more years before Kendrew and Perutz, who by that time were Honorary Visiting Readers (under their mentor, Sir William Lawrence Bragg) at the DFRL, were able to solve the structures of the vitally important

(a)

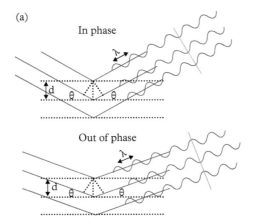

(b)

The Importance of Phase

J. Karle

H. Hauptman

Amplitude of Karle +
Phase of Hauptman

Amplitude of Hauptman +
Phase of Karle

Figure 7.3 *(a) Graphic illustration of the importance of phase in X-ray diffraction studies (courtesy Professor A. M. Glazer); (b) illustration of the importance of phase in the deduction of images from diffraction patterns (by kind permission of the Cambridge Institute of Medical Research, http://www-structmed.cimr.cam.ac.uk/Course/Fourier/Fourier.html).*

(Professor R, Read, Cambridge University)

proteins haemoglobin and myoglobin that transport and store oxygen in the blood of all animals. The reasons for this long delay are given elsewhere,[3] and devolve largely upon finding a means of retrieving the phase, as well as the intensity, of a diffracted X-ray (see Figures 7.3(a) and (b)).

7.4   Two Other Major Contributions to Molecular Biology Made at the DFRL: The First-Ever Structure of an Enzyme and the Study of Viruses

When William Lawrence Bragg, William Henry's son, became Director of the DFRL in 1953, he assembled a group of gifted scientists (largely physicists) to undertake further researches in protein crystallography. The story of the success of this venture has been told both qualitatively and quantitatively, largely by one of the principal protagonist involved in the work, David Phillips (later Lord Phillips of Ellesmere), in his *Scientific American* article in 1966,[8] as well as in the full paper published in *Nature*.[9] A more recent depiction of this seminal advance has been given in Section 6.5.1 of the recent monograph by the present author.[3]

The second example involves the fundamental study of the structure of viruses that Rosalind Franklin and Aaron Klug were pursuing in the Department of Physics at Birkbeck College, University of London, where J. D. Bernal was then Head of Department. Knowing that since the days of William Henry Bragg high-intensity X-rays (rotating anode sources) were available at the DFRL, Bernal sought William Lawrence Bragg's permission for the Rosalind Franklin—Aaron Klug team to use that facility. The team consisted of the research students John T. Finch and Ken C. Holmes (each, in turn to become Fellows of the Royal Society), who, at the time, were Ph.D. students of Rosalind Franklin. These students had access to the high-intensity X-ray sources to investigate the structure of newly discovered viruses. It is fascinating to recall that before Finch could undertake his studies of the human papilloma virus, he had to be first vaccinated by the Salk vaccine, to render him impervious to its infection.[10, 11]

7.5   A Final Word about the Importance of Molecular Biology, its Links with the DFRL, and its Relevance to Modern Medicine

When Sir William Lawrence Bragg became the Fullerian Professor of Chemistry and the Director of the RI and the DFRL, his assembled team of protein crystallographers consisted of David Phillips, a physics graduate from Cardiff University, and two other physicists, A. C. T. North, of King's College, London, and Uli Arndt, of Cambridge University, who had been appointed by Bragg's predecessor, E. N. da C. Andrade. Other key members were Colin Blake and a Ph.D. student who had taken her initial degree in physics, at University College London, Louise Johnson (later Dame Louise). (See Figures 7.4(a) and (b).)

Shortly after graduating, Johnson worked for a few weeks at the Atomic Energy Authority, in Harwell, where she met Uli Arndt. They were both then interested in

Figure 7.4 *(a) Group photograph of the team at the DFRL that solved the three-dimensional structure of lysozyme. From left: Gareth Mair, Colin Blake, Louise Johnson, A. C. T. North, David Phillips, and Raghupathy Sarma. (b) Louise Johnson (see text).*
((a) Courtesy Professor A. C. T. North; (b) Wikipedia open access)

the phenomenon of neutron diffraction. She was very impressed by Arndt's kindness and guidance. He suggested to her that she might wish to join the DFRL to do her Ph.D. She was duly interviewed by Professor Ronald King, who was Sir William Lawrence Bragg's deputy. And so she was accepted as a research student, working mainly with David Phillips. She turned out to be a phenomenally able scientist. After her significant contribution to the solution of the lysozyme structure, she pursued a postdoctoral spell at Yale University. She next joined Phillips at the University of Oxford, where she later became the David Phillips Professor of Molecular Biophysics, and from 2003–08 the Director of Life Sciences at the Diamond Light Source, the UK's national synchrotron facility.

So exceptional was Louise Johnson as a scientist and as a leader of others that, in this book devoted to portraits, personalities, and presentations (at Albemarle Street), it is relevant to cite the passages that Sir Tom Blundell quoted in his *Guardian* obituary of her when she died in 2012. Blundell, an eminent biochemist and crystallographer, wrote a textbook with Johnson, and knew her well as a scientist and colleague. Blundell began his obituary, as follows:

> 'During a special meeting at the Royal Institution in London, in 1965, the first structure of an enzyme, lysozyme, was unveiled by David Phillips. The fully extended polypeptide chain hung down from the high ceiling, coming close to Phillips. In front of him was a much more compact model, defined by X-ray crystallography, of the intricately folded chain. Both represented a protein that was in real life 100 million times as small. Phillips and his colleagues identified a well-defined groove in which evolution had suggestively placed amino acid side chains. The most memorable part of that day was the appearance of Louise Johnson, a young graduate student, who stunned us all by describing how the enzyme bound its substrates (reactants) and selectively cleaned the polysaccharide components of bacterial cell walls, giving rise to its antimicrobial properties, first described by Alexander Fleming in the 1920s. This was the birth of structural enzymology, the beginnings of a continuing investigation of the detailed structures and mechanisms of nature's catalysts.'

At Oxford, as mentioned above, Johnson ultimately became the David Phillips Professor of Molecular Biophysics. Earlier she was a member of Somerville College, before later becoming a Professorial Fellow at Corpus Christi College, Oxford. She was made a Dame in 2003.

Apart from engendering a happy research atmosphere—visible even in the photograph shown in Figure 7.5—this team made major progress in elucidating the structure[8] and mode of action of lysozyme.

As described in the author's recent monograph, members of this team, with special help from a technician, William (Bill) Coates, were the very first of all world crystallographers to elucidate in 1965 in atomic detail the structural features of the enzyme known as lysozyme (which had been discovered by Sir Alexander Fleming in 1922). This is the enzyme that exhibits strong antibacterial action. The structure solved at the DFRL offered a clear picture of how this enzyme carried out its antibacterial activity. It was one of the key turning points of molecular biology (Figure 7.6).

While the above work was in progress, Max Perutz and John Kendrew, on their regular visits to the DFRL, were able to reach their goals: the determination of the detailed structures of myoglobin and haemoglobin, for which they shared the Nobel Prize in Chemistry in 1962.

Figure 7.5 *Group photograph of the scientists, technicians, and administrative staff of the RI and DFRL shortly after Sir William Lawrence Bragg took over his responsibilities there. David Philips is circled left in the middle row, while Bill Coates is circled to his left; Uli Arndt is circled on the back row, while Ronald King is circled to the right of Bragg, also circled, on the front row. Almost half the team were women.*

(By kind permission of Professor A. C. T. North)

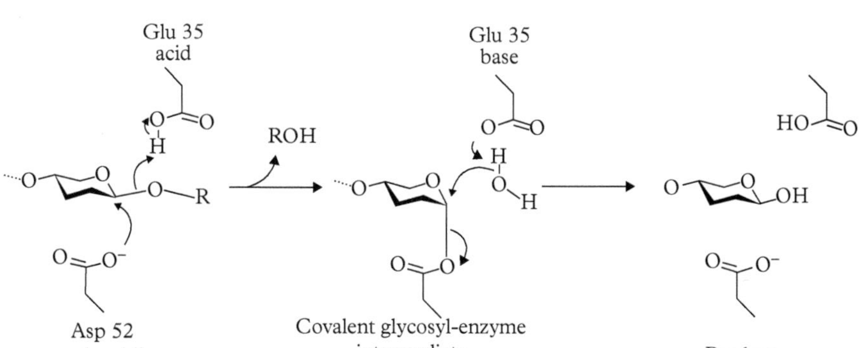

Figure 7.6 *Simplified reaction mechanism of lysozyme in which amino acid residue Asp 52 acts as a nucleophile to form a covalent glycosyl-enzyme intermediate. Amino acid residue Glu 35 acts as both an acid and, subsequently, a base in the reaction.*

(By kind permission of Professor Gideon Davies)

An important practical factor that facilitated both the lysozyme and globin work in London and Cambridge was the very powerful linear automatic X-ray diffractometer, designed by Phillips and Arndt and constructed in the workshop of the DFRL (see the Appendix to this chapter). This instrument, conceived and brought forth entirely in the RI, led the world in its ability to retrieve the intensities of multiple X-ray diffraction spots, needed to solve the structures of the macromolecules of the living world (see Figure 7.1 of the Appendix to this chapter).

It was at this time that there was great jubilation in the RI, in 1965, when the fiftieth anniversary of the award of the Nobel Prize to Sir William Lawrence Bragg was celebrated (Figure 7.7).

Another occasion of note was when both Perutz and Kendrew, with guidance from their mentor William Lawrence Bragg at the RI, and sponsorship from the Medical Research Council, became the key architects of the establishment of the Laboratory of Molecular Biology (LMB) at Cambridge in 1962.

Figure 7.7 *Photograph taken at the RI in 1965 of the fiftieth anniversary of Sir William Lawrence Bragg's Nobel Prize (that was awarded jointly to him and his father in 1915). There are twenty other British Nobel Laureates in this photograph. Dorothy Hodgkin sits behind Lady Bragg, who is at William Lawrence's left.*

(By kind permission of the BBC Photo Library)

Kendrew and Perutz were principally responsible also for establishing the European Molecular Biology Organization (EMBO), which has had a major influence in the rapid, worldwide growth of molecular biology. In due course, the European Molecular Biology Laboratory (EMBL) was established in 1974 in Heidelberg, with Kendrew as its first Director. The EMBL now has outstations in Grenoble, for neutron scattering studies, and in Hamburg, for synchrotron studies of biological macromolecules. It also has a bioinformatics centre in Hinxton, outside Cambridge.

At present (April 2020), twelve Nobel Prizes have been awarded to scientists who were permanently employed at the LMB, while eleven LMB alumni have also received Nobel Prizes, and very many others have had bestowed upon them other distinctions, such as the US Medal of Science and memberships of the foremost academies of the world. The LMB has also become an outstandingly successful body in a commercial sense and in fostering the transfer of technology for the common good (see Chadarevian[14]).

HUMIRA®, an acronym for **HU**man **M**onoclonal Antibody **I**n **R**heumatoid **A**rthritis, is the world's best-selling drug at present. It helps to treat a million patients across the world. The drug Lemtrada®, used to treat multiple sclerosis, as well as the drugs Herceptin® and Opdivo®, for the treatment of breast and lung cancer, respectively, emerged from the work of scientists at the LMB. In the opinion of the recent Head of Medicine at Stanford University, *'the LMB became one of the most successful and influential establishments ever.'* It is to the RI's great credit that its origins can be traced back, step by step, to the perspicacity and influence of William Lawrence Bragg and his cohorts[11,12] at the DFRL, and also to his father, William Henry Bragg, who preceded him there by some thirty years.

APPENDIX 1

The Importance of a Workshop in a Centre of Research Excellence

In their general observations made about centres of excellence, giants such as Rutherford and William Henry Bragg often emphasized how vital it is to have the right machines, instruments, and, above all, expert technicians in a workshop that serves as an integral part of their laboratories. In the account that Max Perutz gave concerning the establishment that he and Kendrew masterminded at the LMB in Cambridge,[15] he emphasized that one of the secrets of their success was the well-equipped workshop, manned by disciplined, dextrous, and gifted technicians, who could be approached readily by the research staff. The research teams dined and drank tea and coffee together with the technical staff.

Not surprisingly, major new advances were made at the LMB, in part because they could create their own unique instruments and techniques. In fact, the DFRL contributed, in a vital manner, to the advances made by Kendrew (and also Perutz) in their pioneering work on the structure of giant protein molecules.

The two papers on myoglobin[9,16] that appeared in 1958 and 1960 had David Phillips of the DFRL as co-author. This is because the LMB group were able to record diffraction data on a revolutionary new diffractometer that was designed and made in the DFRL. This was the single-crystal X-ray linear diffractometer conceived by Phillips, Uli Arndt, and A. C. T. North (see Figure 7.8).

It was a testimony to the scientific vitality of the DFRL that these workers, ably assisted by a team of technicians led by Bill Coats, could leap-frog the majority of X-ray centres in the world by creating this uniquely powerful diffractometer,

Figure 7.8 *The linear diffractometer was largely the creation of David Phillips and Uli Arndt, but A. C. T. North and their technician, Bill Coates, played vital roles in its design, creation, manufacture, and placing of it in the hands of a commercial sales outlet.*
(Courtesy the RI)

which the ingenious John Kendrew took full advantage of in his myoglobin work. With North's skill in using it, Max Perutz also acquired much of his haemoglobin intensity data using this RI diffractomer.[17]

David Phillips, according to the sermon preached at his memorial service in St Margaret's, Westminster, by his former Ph.D. student Louise Johnson, displayed remarkable ability in being able to create the drawings for the diffractometer's construction by the workshop staff, after he had spent a weekend reading the Home University's *'Teach Yourself Engineering Drawing'*. (This was necessary as Ulli Arndt was away in the US at the crucial time.)

There are many lessons to be learnt from the story of the DFRL diffractometer. First, even a small laboratory, equipped with a first-rate workshop and expert technicians, can compete with the finest large centres in the world.[18] Second, and this is related to the first point, a centre of excellence soon fades and loses its reputation once the workshop and technicians are deemed to be dispensable.

To appreciate the full significance of the creation of this diffractometer at the RI, we quote the words of Johnson and Gregory Petsko in a recent *Reflections* article:[21] *'The development of this instrument which was adapted to make multiple simultaneous measurements of intensities. Prior to that time crystallographic data were always collected on X-ray film. The linear diffractometer was the progenitor of all the high-speed electronic data-collection devices (four-circle diffractometers, multi-wire detectors, image plates and CCD detectors) that have revolutionised protein crystallography by improving the quality and the speed of data acquisition by orders of magnitude.'*

APPENDIX 2

What is Sir Lawrence Bragg doing in your Garden?

In an article that Francis Crick contributed to the book that David Phillips and I edited in 1990 to celebrate the centenary of Sir William Lawrence Bragg's birth (*'The Legacy of Sir Lawrence Bragg'*), he told the following story.[19]

> *'Sir Lawrence Bragg was also a keen gardener. When he moved in 1954 from his large house and garden in West Road, Cambridge, to London, to head the Royal Institution in Albemarle Street, he lived in the official apartment at the top of the building. Missing his garden, he arranged that for one afternoon each week, he would hire himself out as a gardener to an unknown lady living in The Boltons, a select inner-London suburb. He respectfully tipped his hat to her and told her his name was Willie. For several months all went well until one day a visitor, glancing out of the window, said to her hostess, "My dear, what is Sir Lawrence Bragg doing in your garden?" I can think of few scientists of his distinction who would do something like this.'*

This story was told to me (and continues to be) whenever I speak of Sir William Lawrence reign as Director of the RI. It is not quite true, however. His daughter

Patience explained after a public talk that I gave at the Athenaeum in London in June 2010. According to Patience (Thomson), it was a friend of hers (in The Boltons) who asked her if she knew of a good gardener that could be hired to work regularly. Patience suggested her father, extolled his quality as a gardener, whom she described also as a versatile handyman, capable of repairing locks and the like. Patience did not describe how elevated a scientist he was, nor did she say that he was the Director of the RI and a Nobel Laureate.

Sir William Lawrence, being the modest, avuncular human being that he was, went along with his daughter's plan. And this is how he was sighted, working in the garden of Patience's friend.

A final word about Sir William Lawrence, whom I never really knew. But three of my technical members of the staff at the RI, who were there in his day, idolized him. In particular, Bill Coates, the general factotum, Bruce Morris, workshop technician, and Jean Conisbee, the person in charge of photography, slides, and illustrations. And no one to my knowledge thought more highly of William Lawrence Bragg than Max Perutz and John Kendrew. Bragg united C. P. Snow's two cultures, according to Perutz[20]

'because his approach to science was an artistic, imaginative one. He thought visually, rather than mathematically, generally in terms of concrete models that can be either static, like the crystal structures, or dynamic…like mobile dislocations in metals. His artistic gifts surfaced in his delicate sketches and water colours, and in his limpid prose.'

Kendrew said[20] of him: *'Basically, he was a puzzle-solver: to him the great fascination was to interpret the complicated diffraction patterns, say of a protein crystal, in terms of its three-dimensional structure…Bragg was one of the last of the classical physicists, who never involved himself much with the ideas of quantum theory or of particle physics.'*

Even after retiring as Director, he used his retirement time at the RI, alongside George Porter, writing beautiful articles and presenting the occasional stimulating general lecture. His *'The Spirit of Science'*[22] was presented to the Royal Society of Edinburgh, and *'What Makes a Scientist?'*[23] was the subject of an RI Discourse. A more complete list of his talks is given in several sources. See refs [24], [25], and [26].

REFERENCES

1. W. H. Bragg, *Lancet*, 21 December **1929**.
2. W. H. Bragg, *Nature*, **1933**, *132*, 11 and 50.
3. J. M. Thomas, *'Architects of Structural Biology: Bragg, Perutz, Kendrew and Hodgkin'*, Oxford University Press, Oxford, **2020**, Chapter 3.
4. Sections 3.4 and 3.5 of ref [3].
5. In his 1938 report for that Foundation, under the heading 'Natural Sciences', Weaver wrote *'Among the studies to which the Foundation is giving support is a sense in a relatively new field which may be called molecular biology, in which delicate modern techniques are now being used to investigate evermore minute details of certain life processes.'*

6. J. D. Bernal, I Fankuchen, and M. F. Perutz, *Nature*, **1938**, *141*, 523.
7. D. Crowfoot and J. D. Bernal, *Nature*, **1934**, *138*, 744.
8. D. C. Phillips, *Scientific American*, **1967**, *215*, 78.
9. C. F. Blake, D. F. Koenig, G. A. Mair, A. C. T. North, D. C. Phillips, and V. R. Sarma, *Nature*, **1965**, *206*, 757.
10. A. Klug, J. T. Finch, and Rosalind E. Franklin, *Nature*, **1957**, *179*, 683.
11. J. T. Finch and A. Klug, *Nature*, **1959**, *183*, 1709.
12. M. F. Perutz, in *'The Legacy of Sir Lawrence Bragg'* (eds J. M. Thomas and Sir David Phillips), **1990**, P. 6.
13. D. C. Phillips, *Biographical Memoirs of Fellows of the Royal Society; Sir Lawrence Bragg*, **1979**, *25*, 1.
14. S. Chadarevian, *Isis*, **2011**, *102*, 601.
15. Ref [3], Chapter 5.
16. J. C. Kendrew, R. S. Dickerson, B. E. Sandberg, R. G. Hart, D. R. Davies, D. C. Phillips, and V. C. Shire, *Nature*, **1960**, *185*, 422.
17. M. F. Perutz, M. S. Rossmann, A. F. Cullis, H. Muirhead, G. Will, and A. C. T. North, *Nature*, **1960**, *185*, 474.
18. My predecessor, George Porter, and his Deputy, Professor David Phillips, was also convinced of the veracity of this claim. He was able, in a twenty-year period, to compete with the best photochemists in the world because of his expert workshop. And so was I, in most of my time, and my Deputy, Professor Richard Catlow, in the fields of solid-state chemistry and heterogeneous catalysts.

    George Porter would often quote the example of Peter D. Mitchell, who worked with two other researchers in a converted mansion in Bodmin, Cornwall, for twenty years to clarify the nature of the synthesis of adenosine triphosphate (ATP). This earned him the Nobel Prize in Chemistry in 1978. Another example of where a small department in a university could revolutionize a technique ahead of the major large universities of the world. Professor E. R. Andrew did this in Bangor, North Wales, when he invented the technique of magic-angle-spinning nuclear magnetic resonance in 1956 (see Chapter 8).
19. J. M. Thomas and D. C. Phillips, *'Selections and Reflections: The Legacy of Sir Lawrence Bragg'*, Science Reviews Ltd, **1990**.
20. J. M. Thomas, *Angew. Chemie. Intl. Ed.*, **2012**, *51*, 12946.
21. Quoted in L. N. Johnson and G. A Petsko, *Trends Biochem. Sci.*, **2009**, *24*, 287.
22. W. L. Bragg, *Proc. Roy. Soc. Edinburgh*, **1967**, *67*, 303.
23. W. L. Bragg, *Proc. Roy. Inst. Great Brit.* **1969**, *42*, 397.
24. Ref [3], Chapters 6 and 8.
25. J. M. Thomas, *Notes and Records of the RS*, **2011**, *65*, 163.
26. D. C. Phillips, opening chapter of ref [19].

8

Modern Diagnostic Medicine: Memorable Discourse in 1986 by Raymond Andrew on MRI

8.1 Introduction: How Untrammelled Curiosity Leads to Major Technological Advance–the Arrival of MRI and PET[1]

Members of the general public sometimes forget that major scientific applications, especially those used in hospitals, often arise as a result of the untrammelled intellectual curiosity of individual research scientists, who, in their investigations, are solely concerned with knowing more about the fundamental properties of matter. This monograph, and others, demonstrate *inter alia* that Faraday's ideas about the common origin of the forces of nature led him to discover electromagnetic induction, to pursue electrochemistry, and also to initiate magneto-optics.

Faraday, from the beginning of his experimental work, was also acutely aware of how a particular discovery, however minor (in the context of his major advances), could possibly be put to good practical use. Thus, in his early paper in 1825[1] entitled *'On New Compounds of Carbon and Hydrogen and Certain Other Products Obtained By The Decomposition of Oil by Heat'*[1], in which he describes the discovery and identity of benzene, he ends on a very practical speculation, as follows:

> *'It is possible that, at some future time, when we better understand the minute changes which will take place during the decomposition of oil, fat, and other substances by heat, and have more command of the process, that the substance, among others, may furnish the fuel for a lamp, which remaining a fluid at the pressure of two or three atmospheres, but becoming a vapour at less pressure, shall possess all the advantages of a gas lamp, without involving the necessity of high pressure.'*

---

[1] In writing this chapter, I received invaluable guidance and help from Professors Terry Jones (Davis, California), Adrian Dixon (Cambridge), and J. D. Pickard (Cambridge).

Albemarle Street: Portraits, Personalities, and Presentations at the Royal Institution. John Meurig Thomas, Oxford University Press. © Sir John Meurig Thomas 2021.
DOI: 10.1093/oso/9780192898005.003.0008

It was Faraday's constant pondering of what the relationship was between electricity and magnetism, and whether they could be interconverted, that led to his discovery of electromagnetic induction (and hence the dynamo and transformer), as described in Chapter 4.

Almost all the major hospitals of the world nowadays use two medical techniques that have been, and continue to be, of enormous importance in the diagnosis of disease and in the determination of the functioning of the healthy components of the human body, especially the brain. The first of these is magnetic resonance imaging (MRI), the origin of which goes back to 1946 when two independent groups of physicists—one in Harvard,[2] the other in Stanford[3]—were concerned about the magnetic properties of the nuclei of certain atoms. This led to the discovery of a branch of spectroscopy, called nuclear magnetic resonance (NMR), that has proved of inestimable importance in the determination of the structure of new materials, from small and giant molecules to solid materials such as organic polymers, proteins, carbohydrates, as well as ceramics, silicates, and other minerals.

Apart from the vitally important role that MRI has played—in being superior to X-ray imaging (as explained below)—another major imaging technique is that known as positron emission tomography (PET). Positrons were not known until their existence was postulated by P. A. M. Dirac at Cambridge University in a period of intense activity by him as a theoretical physicist in 1928. His quantum mechanical work convinced him that a hitherto unknown particle, the positron, had to exist. He claimed that it would be the same size and mass as the electron, but bearing a positive charge. It was the first inkling of the existence of antimatter. An experimentalist named C. D. Anderson, in 1932, working in the California Institute of Technology, detected the presence of positrons in cosmic rays. Nowadays, positrons can be produced routinely in the laboratory—either by irradiating a metal, such as gold, with ultra intense high-power lasers—or much more conveniently by 'picking them up', as it were, as the decay products of certain unstable, radioactive isotopes. Two very important isotopes that emit positrons, one of carbon and the other of fluorine, are used widely in this so-called 'nuclear medicine' method of imaging the body, and investigating the nature and progress of any disease that it may have. We describe how this is done in Section 8.8.6 below.

8.2   The Principles of NMR

In late 1940s, the nuclei of atoms were objects of great mystery. In addition to being characterized by their mass and their electrical charge, nuclei were also known to possess two other properties: magnetism and spin angular momentum (abbreviated 'spin'). The property of magnetism means that we can think of the nuclei as being tiny bar magnets. The concept of spin is illustrated by the Earth which spins about an axis through its centre once every twenty-four hours. But, in 1949, quantitative knowledge of these two properties was available for only a few nuclear species.

The invention of NMR made possible a simple yet powerful method of measuring the magnetism and spin. In its simplest form, it measures the ratio of the magnetism to the spin angular momentum, the so-called gyromagnetic ratio.[4]

The set-up shown in Figure 8.1 is a schematic diagram of the essence of an NMR experiment.[5] (Bear in mind the similarity of this set-up to Faraday's electromagnetic induction.)

To study a particular nucleus, we find some substance that contains that nucleus. For example, ^{13}C (the proper scientific notation for the carbon isotope of atomic mass 13) is found in graphite, in diamonds, in carbon tetrachloride, in ethyl alcohol, in nylon, in petrol, etc. Pick whatever substance is convenient, fill a small test tube with it, and put it into a small electrical coil. Connect the coil to a radio transmitter and to a radio receiver. Lastly, insert the coil into the jaws of a large magnet. If the strength of the magnetic field produced by the magnet is called B, the two groups (at Harvard and Stanford) showed that one can detect the magnetism of the nuclei by tuning the radio transmitter and the radio receiver to a frequency, f, given by the relationship:

$$f = \frac{1}{2\pi}\gamma B \tag{1}$$

The strength of the signal is proportional to the number of nuclei in the sample that produces it. If we know B (how strong a magnet we are using), we can measure γ by finding the frequency at which the nucleus will give a signal. You can think of the task as trying to locate where each type of nucleus is located on the 'nuclear radio dial'.

In 1951,[6] a Stanford University group reported that the NMR spectrum of ethyl alcohol, formula CH_3CH_2OH, shows it to consist of a CH_3 group attached to a CH_2 group, and in then comes OH group. Figure 8.2 shows the hydrogen NMR signal. Since the NMR signal is proportionate to the number of nuclei in the sample, we immediately understand the nature of the signals in Figure 8.2.

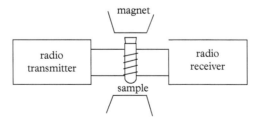

Figure 8.1 *Schematic diagram of a nuclear resonance apparatus.*

In 1953, Gutowski and co-workers in Illinois published a very important paper.[7] This demonstrated that the resonance frequency of a proton in a given molecular fragment (the so-called chemical shift when referred to some substance that is chosen as a reference standard) is characteristic of that fragment, no matter what the structure of the rest of the molecule is (Figure 8.3). This important paper

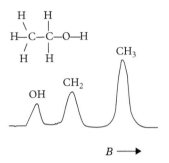

Figure 8.2 *The proton NMR spectrum (schematic) of the molecule CH_3CH_2OH, ethyl alcohol, observed by Arnold, Dharmatti, and Packard[6] (nuclear absorption versus magnetic field strength B).*

Hydrogen Chemical Shifts   Meyer, Saika, and Gutowsky (1953)

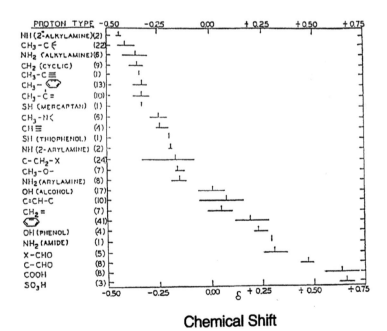

Chemical Shift

Figure 8.3 *The shift in frequency (in parts per million of frequency) of hydrogen nuclei in various families of chemical compounds, measured by Gutowski et al.[7]*
(Courtesy Professor C. P. Slichter[5])

demonstrated the power of NMR; it demonstrated that particular molecular fragments have a unique signature.

Several other technical advances in NMR spectroscopy were made in subsequent years, which, in part, took advantage of instrumental and computational developments. Individuals such as Ernst, Freeman, Wurtrich, and Jenner made significant contributions that enabled NMR spectra to be collected rapidly using Fourier transforms of free-induction decays as well as recording two-dimensional spectra—see the review by Professor C. P. Slichter for further details.[5]

8.3   Raymond Andrew and the Magic Angle

First, a word about Raymond Andrew. After graduating from Christ's College, Cambridge (where one of his physical chemistry teachers was C. P. Snow), he proceeded to work on superconductivity—interrupted by wartime work on radar—and then proceeded to Harvard to join the group of E. M. Purcell. On his return to the UK, he carried out work on NMR, first at St Andrews, Scotland, and then he became the Professor of Physics at the University College of North Wales, in Bangor. It was there that he carried out a most elegant experiment, with his collaborators, and which gave rise to magic-angle-spinning NMR (MASNMR).

Whereas NMR spectra of liquid and gaseous samples are sharp and enable chemical shift values to be readily recorded, for solids, NMR spectra are very broad, because of the existence of interactions which, in liquid and gases, are averaged by the rapid thermal motion of molecules and ions. In solids, such motion is lacking. For example, the solution H NMR spectrum of camphor (see Figure 8.4(a)), a compound known since antiquity and one which played an important role in the development of organic chemistry, contains a wealth of information.

The parameters derived from it (positions, widths, intensities, the multiplicities of lines) provide unique information on the structure, conformation, and molecular motion. By contrast, the NMR spectrum of the solid compound (see Figure 8.4(b)) consists of a single broad hump, which conceals information of interest to a chemist. High-resolution spectra (i.e., those that enable magnetically non-equivalent nuclei of the same spin species to be resolved as individual lines) can be obtained only when the anisotropic interactivity giving rise to line broadening are substantially reduced.

Without going into the detail of why the lines of nuclei in solids are so broad—they depend on dipolar interactions with neighbouring magnetic nuclei (see Figure 8.5)—we can indicate why NMR lines are broad in solids. What Andrew realized (and a US worker, Lowe, independently came to the same conclusion) was that if the sample was rapidly spun at an angle of $54°44^1$—the magic angle—

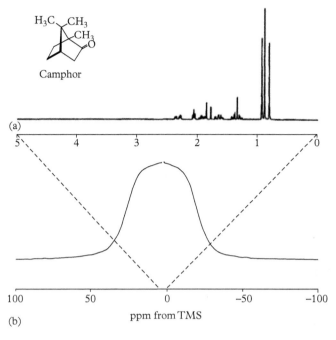

Figure 8.4   *^1H NMR spectra of camphor (a) dissolved in $CDCl_3$; (b) stationary solid sample. Note the forty-fold scale difference between the two spectra.*[9]

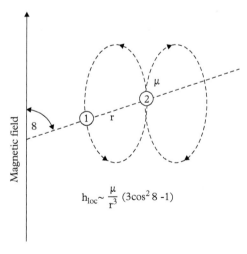

Figure 8.5   *Nucleus 2 produces a local magnetic field at the site of nucleus 1. When $\theta = 54°44'$, then $3\cos^2\theta-1 = 0$ and $h_{loc} = 0. \mu$ is the nuclear magnetic moment.*

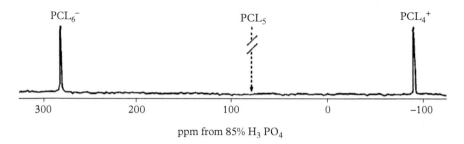

Figure 8.6 *The ^{31}P MASNMR spectrum of powdered PCl$_5$ is composed of two signals of equal intensity.[11] The arrow indicates the position of the single resonance from PCl$_5$ in solution. (Courtesy Raymond Andrew)*

then a sharp NMR line resulted, just as if the solid had become a liquid. It is the occurrence of a term (3cos$^2\theta$-1) (see Figure 8.5) that leads θ = 54°44^1 to yield a zero term that is the key to this situation.

Andrew, with his colleagues, and a small, but competent workshop in Bangor, were able to build a set-up that enabled solid samples to be mounted in a turbine spinner (capable of 5,000 revolutions per second) and inclined at the magic angle, 54°44^1, to the magnetic field.

The proof that Andrew et al. offered for the validity of this approach to retrieving a sharp line for magnetic nuclei in solids was an elegant one. It involved the pentachloride of phosphorous PCl$_5$. The substance in solution is a discrete molecule. But in the solid-state it exists as aggregates of ionic pairs: PCl$_4^+$ and PCl$_6^-$. When Andrew et al.[8][11] subjected PCl$_5$ in the solid-state to MASNMR, the result was as shown in Figure 8.6.

Numerous applications have been made of MASNMR in both the organic and inorganic domain, and especially in zeolitic heterogeneous catalysis—see refs [9], [12], and [13].

Andrew moved, in 1964, from Bangor to the Department of Physics, University of Nottingham. He continued to carry out outstanding work there. His group and two others in the same department were, in effect, competing to exploit MRI, which we describe next, using paragraphs from Andrew's Discourse at the RI and the work done at about the same time at Champaign, Illinois. After nineteen years in Nottingham, where he developed the MRI scanner, Andrew left Nottingham to go to the University of Florida at Gainesville, which is where he did his most notable MRI work. At Nottingham, his former colleague Peter Mansfield made outstanding developments in MRI. Both he and Paul Lauterbur,[14] of Illinois, were later to win the Nobel Prize for Physics, after Andrew had passed away.

8.4 The Discourse at the RI on MRI by Raymond Andrew in 1986

The following was Andrew's opening statement at the RI:. *'It is of tremendous importance to the physician to be able to see inside the human body to enable him to deal with diseases either by medical or surgical treatment. One of the best ways of seeing inside the human body is by using X-rays and, particularly with the development of computerised tomographic (CT) X-ray scanning, extremely good pictures or images can be obtained. But X-rays do have one serious disadvantage: they are an ionising radiation and so they can do harm in the course of doing good.*

'By contrast, obtaining images by magnetic resonance has the great virtue that it does not use ionising radiation and is therefore inherently safer. Quite apart from this freedom of hazard, magnetic resonance images offer additional information through tissue discrimination and pathological discrimination, which arises from differences of so-called relaxation time, a parameter not available in X-ray scanning. Furthermore, one can obtain images in any desired plane, transverse, sagittal or coronal, with equal ease. With CT X-ray scanners, one can only obtain direct images in the transverse plane.'

Andrew proceeded to remind the RI audience that each human being contains some 10^{28} hydrogen atoms, and we are all in the Earth's magnetic field of 0.5 Gauss (or 50 µ Tesla). Consequently, all 10^{28} hydrogen nuclei in our bodies are busy processing (see Figure 8.7) at about 2 kHz (2,000 per second) in the Earth's field. In Figure 8.7, the notion of a hydrogen nucleus processing in a magnetic field (at the so-called Larmor frequency) is shown.

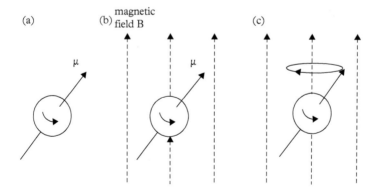

Figure 8.7 *(a) A hydrogen nucleus (proton) with a magnetic moment µ. (b) The hydrogen nucleus in a magnetic field. (c) The hydrogen nucleus processes in the magnetic field at the Larmor frequency.*[16]

(Courtesy Raymond Andrew)

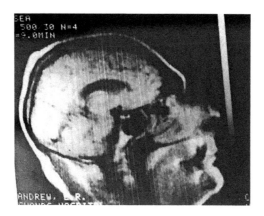

Figure 8.8 *MRI image of a thin sagittal midline section of Raymond Andrew's head.*
(Courtesy Raymond Andrew)

Based on the physics that he explained from Figure 8.7 onward, Andrew showed an image of a thin sagittal section of his own head (see Figure 8.8).

Such images as this had slowly been improving from the early1970s onward, and a great debt is owed to Paul Lauterbur, along with others, for this advance in medicine. As explained by Slichter,[5] the fundamental idea is to put the object one wishes to image in a magnetic field that varies in a known and controllable manner with position. *'Suppose one were studying the resonance of a particular species of nuclei in the object (e.g., that of hydrogen atoms). Since the resonance frequency is proportional to the strength of the magnetic field (Equation 1), different parts of the object would have different resonance frequencies. One could therefore associate a given resonance frequency with a given position. The size of the resonance signals at each frequency tells how many nuclei are at that frequency, hence at the positions at which the corresponding magnetic field occurs.'*

The concept is illustrated in Figure 8.9. Here is a glass of water. We record the magnetic resonance of the hydrogen nuclei in the glass. If the glass is in a uniform magnetic field, nuclei in all parts of the glass will have the same resonant frequency. But if we arrange the magnetic field to be high at the top of the glass and low at the bottom, the water at the top of the glass will have a resonance at a higher frequency than those at the bottom. The resonance would then span a range in frequencies and each frequency would correspond to a unique vertical position.

If we drank some water from the top of the glass, the frequency at the top of the water in the glass would be lower, and the absorption would not go quite as high in frequency. If we know how frequency varied with position we could deduce the new level of the water. If one had used a martini glass instead of an ordinary water

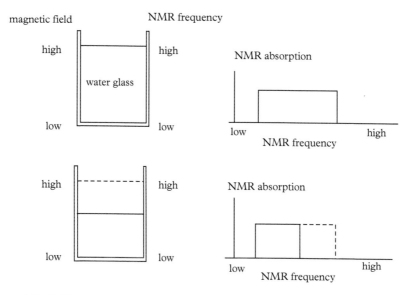

Figure 8.9 *Full (upper left) and partially full (lower left) glasses of water in a magnetic field that is higher at the top of the glass than at the bottom. The corresponding NMR absorption curves of the water (right upper and lower) show that the higher frequency part of the NMR absorption of the full glass is not present in the absorption signal of the partially full glass.*[5]
(By kind permission of the American Philosophical Society)

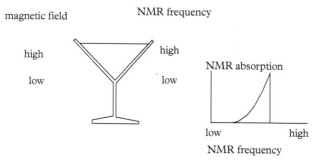

Figure 8.10 *On the left, a martini glass in a magnetic field that is stronger at the top of the glass than at the bottom. On the right, the hydrogen nuclear absorption spectrum of liquids in the glass. The conical shape of the martini glass causes the lower frequency portion of the NMR absorption to be weaker than the high frequency, in contrast to the signals in Figure 8.9.*[5]
(By kind permission of the American Philosophical Society)

glass, the resonance absorption signal would be of a different value, as shown in Figure 8.10. Because the glass is shaped like a cone, the nature of the signal is very different. One can easily deduce what the glass container is shaped like from the nature of the spectrum.

The basic idea of image formation is simple enough, but translating the idea into a practical system involves major engineering and design work. The first commercial imaging apparatus did not appear until about 1983. At about that time, Raymond Andrew's head (Figure 8.8) was imaged in the University of Florida. It took nine minutes to make.[16]

8.5   A Word About Functional MRI

When an area of the brain is in use, blood flow to that region also increases, and so functional MRI (fMRI) is used to measure brain activity by detecting change associated with blood flow. This technique relies on the fact that neuronal activation is associated with cerebral blood flow. The primary principal that governs fMRI is blood-oxygen-level-dependent (BOLD) contrast. fMRI is based, according to one of its expert practitioners, J. D. Pickard, on the curious phenomenon of the BOLD effect. Curious because the increase in blood supply to active parts of the brain exceeds that required to meet the increased metabolic demand. Deoxyhaemoglobin concentration falls and oxyhaemoglobin concentration rises. fMRI detects the difference in magnetic properties.

It was S. Ogawa at the Bell Laboratories in 1990 who pioneered fMRI, which he did by mapping neural activity in the brain (or spinal cord) of humans and animals by imaging the change in blood flow related to energy use by brain cells.[17] The technique of fMRI has been one of the dominant ones for mapping brain activity since it was first introduced (but see Section 8.6); one of the reasons for this is that it does not require the subjects (patients) to undergo injections or surgery, or to ingest unpleasant substances, or to be exposed to ionizing radioactive surgery.

While it is extensively used, it is recognized that sources of noise in the measurements can interfere. In clinical work, where fMRI is popular, a series of processing steps must be performed in the acquired images before the actual statistical search for task-related situations can begin.

A particularly interesting example of the use of fMRI is seen in the work of A. Owen and Pickard and their colleagues in 2006, when they reported on detecting awareness in the vegetative state.[18] The vegetative state is one of the least understood and, according to Owen et al.,[18] 'most ethically troublesome conditions' in modern medicine. A patient may, for example, emerge from coma and appear to be awake, but shows no signs of awareness. What Owen and colleagues did is summarized in the set of four fMRI images shown in Figure 8.11.

The researchers carried out a series of fMRI measurements on a patient that was in a vegetative state. The patient was given spoken instructions to perform two mental imagery tasks at specific points during the MRI scan. One task involved imagining playing a game of tennis and the other involved imagining

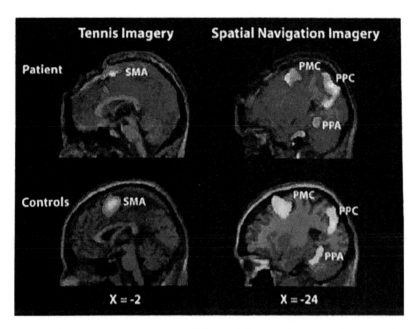

Figure 8.11 *Supplementary motor area (SMA) activity was observed during tennis imagery (see text) in the patient and a group of twelve healthy volunteers (controls). The authors detected parahippocampal gyrus (PPA), posterior parietal cortex (PPC), and lateral premotor cortex (PMC) activity while the patient and the same group of volunteers imagined moving around a house. (By kind permission of Professor J. D. Pickard, A. Owen, and HHAS)*

visiting all the rooms of the patient's home, starting from the front door. During the periods when the patient was asked to imagine playing tennis, significant activity was observed in the supplementary motor areas. In contrast, when the patient was asked to imagine walking through her home, significant activity was observed in the parahippocampal gyrus, the posterior parietal cortex, and the lateral premotor cortex, as indicated in Figure 8.11—see legend, in particular. The neural responses were indistinguishable from those observed in healthy volunteers performing the same imaginary tasks in the MRI scanner.

These results confirm that, despite fulfilling the clinical criteria for a diagnosis of vegetative state, this patient retained the ability to understand spoken commands and to respond to them through her brain activity, rather than through speech or movement. *'Moreover'*, to quote the authors of the paper, *'her decision to cooperate with the authors by imagining particular tasks when asked to do so represents a clear act of intention, which confirmed beyond any doubt that she was consciously aware of herself and her surroundings.'*

8.6   PET and Transformative Advances in Medicine

No Discourse has yet, to my knowledge, been given at the RI which focuses on a technique that must also be mentioned alongside MRI and fMRI, since it is able to interrogate the brain, the central nervous system, and other regions of the human body non-invasively. Very recently it is able to study the whole body at once, non-invasively.

PET is the most sensitive and specific technique for imaging molecular interactions and pathways in the human body. It is an advanced form of molecular imaging as opposed to anatomical imaging undertaken by X-ray- or MRI-based procedures. PET rests upon administering minute amounts of a molecule labelled with a positron-emitting radio nuclide. The radiolabelled molecule is selected to act as a tracer to provide a biomarker of a biological process, e.g., a specific binding site or pathways in tissue.

As mentioned above, a popular radioactive material used in PET is fluoro-2-deoxy-D-glucose (termed [^{18}F] FDG) (see Figure 8.12).

This modified molecule of glucose is very useful in identifying the location (and progress) of tumours. This molecule is rapidly consumed at a tumour, and hence it signifies the precise location and degree of that tumour.

When the radio nuclide undergoes nuclear decay, it emits a positron. The emitted positron travels a short distance in the tissue before it is captured by a negatively charged electron. As a result of the capture, both particles annihilate, and a pair of photons are emitted at approximately 180 degrees to each other. Hence, emerging from the body are these coincident pairs of photons. By using radiation detectors that work in coincidence, it is possible to identify lines through the body where this capture of the positron occurred. By recording many lines through the body, it is possible to have sufficient data to reconstruct the three-dimensional distribution of the tracer molecule. This distribution is often shown as tomographic cross sections through the body. This emission tomography is the reverse of X-ray computerized tomography (CT), which reconstructs the density of tissue as measured from the attenuation of the transmission of photons through the body. Figure 8.13 shows a PET scan of a cancer patient with lymphoma scanned with a PET biomarker of a glucose analogue deoxyglucose radiolabelled with

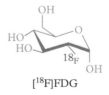

[^{18}F]FDG

Figure 8.12   *Conformational structure of fluoro-2-deoxy-D-glucose (termed [^{18}F] FDG) widely used in positron emission tomography.*

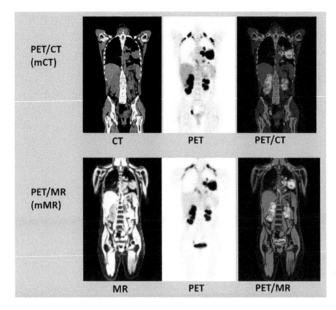

Figure 8.13 *Comparison of PET scans of a cancer patient and corresponding CT scan (top) and MRI scan (bottom). Note (mCT) and (mMR) refer to the comparison of the molecular (m), i.e., PET scan with the CT and MR respectively.*
(Courtesy D. M. Townsend, Clinical Imaging Research Centre, Singapore)

positron-emitting fluorine-18 (see below). The PET scan has been combined with X-ray CT and MRI to provide the complimentary anatomical scans.

The high specificity of the PET scan arises from the fact that a specific molecule, administered as a tracer, provides the signal as opposed to the CT and MRI scans, which respectively image tissue density and water in different configurations. The superior sensitivity of PET rests on the radioactive signal being recorded with very little background noise, and detection of the emitted photons is undertaken electronically by coincidence counting surveying large volumes of the body.

An invaluable outline of the essence of PET in modern diagnostic medicine has been provided by T. Jones.[19] He explains why ^{18}F, ^{15}O, and ^{11}C are popular radionuclides to use as positron emitters and how they are produced using a cyclotron.

Figure 8.13 shows a PET scan of a cancer patient with lymphoma scanned with a PET biomarker of [^{18}F] FDG. Many human tumours metabolize excessive amounts of glucose, and the dark hot-spot areas delineate the local growth of the lymphoma. It is common practice to combine PET scans with X-ray CT scans, and also with MRI imaging, so as to provide a complementary anatomical scan to the molecular ones (produced by PET).

8.6.1 Clinical Research and Healthcare Applications[20]

Initially, when PET was used for medical diagnoses it was mainly applied to the study of brain processes; and it succeeded to delineate the cartography of the brain.[21] Gradually, however, attention turned to cardiology and then oncology. In due course, PET became a crucial technique in the procedures of healthcare. An early account of how PET was used to investigate brain disorders was given by Jones and Rabiner.[22] In cardiology, regional imaging of myocardium—the muscle of the heart—blood flow and metabolism have been the predominant PET procedures. And, in oncology, as implied earlier, use of the fluoroglucose, radiolabeled molecule has been extensive.

The introduction of total-body PET scanning[23] is destined to provide examples of transformative clinical research (see Figure 8.14[23]).

Thus, the work of Badawi et al.,[23] using the so-called EXPLORER[24] total-body PET scanner, constitutes the first medical-imaging scanner of any kind that is capable of capturing three-dimensional images of the entire human body at the same time. It enables studies of the kinetics of an injected radiotracer throughout the entire body. Diagnostic-quality scans using requisition times of one minute or less can be obtained. The EXPLORER total-body scanner has the ability to image [^{18}F] FDG distribution for up to ten hours (i.e., some five half-lives of the radio nucleotide) after injection, thereby expanding greatly the knowledge that can be retrieved using PET.

It is remarkable that from 1986, when Raymond Andrew astonished his RI audience, when he showed the MRI image of his own head (Figure 8.8), it is now possible to follow chemical and metabolic changes as well as simultaneous anatomical changes (by MRI and CT) of whole bodies using PET.

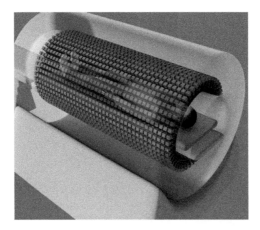

Figure 8.14 *Configuration of the EXPLORER total-body PET scanner covering nearly 200 cm of the axial length of the body.[23] (S. R. Cherry, R. D. Badawi, J. S. Karp, W. W. Moses, P. Rice, and T. Jones, Sci. Travel Med., eaaf6199, 2017.)*

APPENDIX

A Pictorial Analogue of CT

In the world of diagnostic medicine these days, frequent reference is made to tomography and tomograms. Ever since CT scans were introduced, the layman has been exposed to that term almost incessantly whenever he or she enters a hospital. This appendix aims to convey to the reader the essence of three-dimensional imaging in a general context.

The word tomogram itself comes from the Greek, and it means 'slice' or 'slicing'. If one turns to the field of electron microscopy it is easier to understand, in pictorial terms, how tomography enables three-dimensional images to be constructed from a series of two-dimensional slices. Klug worked out a tomographic approach to reconstructing a three-dimensional image from a series of two-dimensional ones recorded by electron microscopy.[25] He did this by what is called Fourier reconstruction. The essence of this method can, however, be readily appreciated by reference to the similar method of back projection summarized in Figure 8.15, and much used by both W. Baumeister[26] and P. A. Midgley[27] and their respective co-workers.

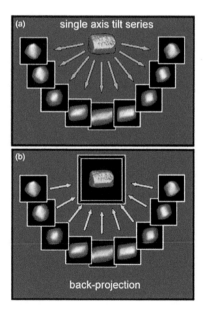

Figure 8.15 *Schematic diagram illustrating the principles of tomographic reconstruction using the so-called back-projection method. In (a) a series of images is recorded at successive tilts. These images are back projected in (b) along their original tilt directions into a three-dimensional object space. The overlap of all the back-projections defines the reconstructed object.[27]*
(Courtesy Royal Society of Chemistry)

REFERENCES

1. M. Faraday, *Phil. Trans. R. Soc.*, **1825**.
2. E. M. Purcell, H. C. Torrey, and R. V. Pound, *Phys. Rev.*, **1946**, *69*, 37.
3. F. Bloch, W. W. Hansen, and M. Packard, *Phys. Dev.*, **1946**, *69*, 127.
4. Sometimes this is called the 'magnetogyric' ratio.
5. C. P. Slichter, *Proc. Amer. Phil. Soc.*, **1998**, *142*, 533.
6. J. J. Arnold, S. S. Dharnate, and M. U. Packard, *J. Chem. Phys.*, **1951**, *19*, 507.
7. L. H. Meyer, A. Saika, and H. S. Gutowski, *J. Amer. Chem. Soc.*, **1953**, *76*, 4567.
8. E. R. Andrew, A. Bradbury, and R. G. Eades, *Nature*, **1958**, *182*, 1659.
9. J. Klinowski and J. M. Thomas, *Endeavour*, **1986**, *10*, 2.
10. I. J. Lowe, *Phys. Rev. Lett.*, **1959**, *2*, 285.
11. E. R. Andrew, A. Bradbury, R. G. Eades, and G. J. Jenks, *Nature*, **1960**, *188*, 1096.
12. S. Ramdas, J. M. Thomas, K. Klinowski, C. A. Fyfe, and J. S. Hartman, *Nature*, **1981**, *292*, 228.
13. E. Lippmaa, M. Magi, A. Samason, M. Tarmuk, and G. Engleharat, *J. Am. Chem. Soc.*, **1981**, *103*, 4992.
14. P. C. Lauterbur, *Nature*, **1973**, *242*, 190.
15. P. Mansfield and P. K. Granell, *J. Phys.*, **1973**, *C6*, L422.
16. E. R. Andrew, *Proc. Roy. Inst. Great Brit.*
17. S. Ogaura, F. M. Lee, and A. S. Nayak, *Magnetic Resonance in Medicine*, **1990**, *14*, 68.
18. A. M. Owen, M. R. Coleman, M. Baly, M. H. Davis, S. Laurleys, and J. D. Pickard, *Science*, **2006**, *313*, 1402.
19. T. Jones, *Eur. J. Nucl. Med.*, **1996**, *23*, 207.
20. I am indebted to Professor T. Jones for providing the information contained in this section.
21. B. Holford, *Chem. Eng. News*, **2008**, September, 13.
22. T. Jones and E. A. Rabiner, *Cerebral Blood Flow and Metabolism*, **2012**, *32*, 1426.
23. R. D. Badawi, H. Shi, P. Hu, S. Chen, Y. Xu, P. M. Price, Y. Ding, B. H. Spencer, L. Nando, W. Liu, J. Bao, T. Jones, H. Li, and S. R. Cherry, *J. Nucl. Med.*, **2019**, *60*, 299.
24. The EXPLORER total-body PET scanner has been produced by a consortium of experts at: (a) the University of California, Davis; (b) Zhangzhou Hospital, Fudan University, Shanghai, China; and (c) Imperial College London, UK.
25. A. Klug, 'Nobel Lecture, 1982', *Angew. Chemie. Intl. Ed.*, **1983**, *22*, 565.
26. W. Baumeister, *Curr. Opin. Struct. Biol.*, **2002**, *12*, 679.
27. P. A. Midgley, E. P. W. Ward, A. B. Hangria, and J. M. Thomas, *Chem. Soc. Rev.*, **2007**, *36*, 1477.

9

Egyptomania at the RI: Howard Carter's Discourse on the Tomb of Tut-Ank-Amun from Ante-Room to Burial Chamber

9.1 Introduction

When, in 1798, Napoleon invaded Egypt, he set in train a sequence of intense activities politically, culturally, scientifically, linguistically, and archeologically. From the early nineteenth century onward, people in most countries were curious to know more about the ancient kingdom of the Egyptians, whose civilization had flourished for six millennia.

Even as early as 1826, Faraday and his colleagues arranged for a remarkable, Milan-born doctor, Augustus Bozzi Glanville, to present a Discourse at the Royal Institution (RI) on the subject of *'Mummies'*. Glanville was one of the leading physicians in London and a keen writer. In 1825, he had lectured to the Royal Society on his work as one of the first experts to carry out a medical autopsy on an Ancient Egyptian mummy. He later repeated this lecture-demonstration at the RI, and later donated both material he used and his discoveries to the British Museum.

In 1834, a London-based Egyptologist, John Davidson, gave a Discourse on *'The Pyramids of Egypt'*, special attention being given to the Giza complex consisting of the pyramids of the Pharaohs Khufu, Khafre, and Menkaure,[1] built in the Third Dynasty (2635–1700 BC)—see Figure 9.1. Two years later, another medical doctor, Thomas Joseph Pettigrew, described to members of the RI—with an actual example—*'The Opening of an Egyptian Mummy'*.[2]

In 1841, shortly before the time of Jean-François Champollion's early death in Paris, an English expert, Samuel Birch, described at the RI *'The Hieroglyphics of the Egyptians'*, a topic that had occupied the thoughts of Thomas Young (Chapters 1 and 11) nearly thirty years earlier. Like Young, Champollion had an astonishing ability to master a wide range of oriental languages. By the age of fourteen he was

Albemarle Street: Portraits, Personalities, and Presentations at the Royal Institution. John Meurig Thomas, Oxford University Press. © Sir John Meurig Thomas 2021.
DOI: 10.1093/oso/9780192898005.003.0009

Figure 9.1 *The Giza complex of pyramids in Cairo: Khufu, Khafre, and Menkaure.*
(Courtesy Cairo Museum and Dr Zaki Iskander)

fluent in Latin, Greek, Ethiopic, Arabic, Syniac, and Chaldean. Later, he mastered Coptic, which proved invaluable to him as an Egyptologist. Many of these languages, as described earlier, Young also mastered, but not Coptic.

9.2 The Colourful Piazzi Smyth

Intermittently, after Birch's Discourse, an ever-increasing fascination led to numerous others at the RI on the subject of Ancient Egypt. A particularly dramatic one was given by one of the most colourful personalities in the world of British science in the nineteenth century, Charles Piazzi Smyth (see Figure 9.2), who was for forty years Astronomer Royal of Scotland.

Apart from being an astronomer[3] and a spectroscopist, he was also a pioneer photographer, an artist, a meteorologist, a metrologist, a traveller, a writer, and an enthusiastic pyramidologist, believing that there was some mystical significance to the dimensions of the Great Pyramid of Khufu at Giza.

Piazzi Smyth's rather bizarre views on the pseudo-religious significance of the dimensions of Khufu, now widely ridiculed, are described briefly in Appendix 1 of this chapter.

Figure 9.2   *Charles Piazzi Smyth, Astronomer Royal of Scotland, who entertained bizarre views about the great pyramid—see text and Appendix 1.*
(Wikipedia open access)

Figure 9.3   *Sir Flinders Petrie, the first Professor of Egyptology in the UK (at University College, London).*
(Wikipedia open access)

Other Egyptologists in subsequent years have performed at the RI. Thus, the great Flinders Petrie, the first Professor of Egyptology in the UK (University College London), and the man who hired the young Howard Carter to assist him in the important excavation of Tel El Amarna, where the Pharaoh Akhenaten and his Co-Regent Nefertiti were domiciled, gave an RI Discourse on Egyptian jewellery in February 1916.

In 1892, Petrie went to study the Great Pyramid of Khufu and concluded that all the claims and theories of Piazzi Smyth were erroneous. He later made monumental discoveries in Egypt.

Insofar as the jewels of Ancient Egypt are concerned, minerals were extensively used, as explained by the eminent Egyptologist Cyril Aldred (see Section 9.4). According to Aldred, the Egyptians selected minerals for their colour and polish, rather than for their rarity, refractive power, or brilliance. The amuletic pectorals worn as jewellery in those days as protection from mysterious hostile forces consisted, as Aldred has reminded us, of various stones such as red carnelion, green turquoise or green beryl, and blue lapis lazuli, which preserved within themselves the colour of lifeblood, the fresh green of springing vegetation, or the blue of water and the holy sky-realm.

9.3 Howard Carter and his Discourse

Although born in Kensington, London, Howard Carter (Figure 9.4) spent most of his childhood and youth in the market town of Swaffham, in Norfolk.[4] At the age of seventeen, he was fortunate to obtain a post in Egypt, and in 1892 he worked under Flinders Petrie, who recognized his exceptional skills as a draftsman and painter, who could faithfully copy the symbols and illustrations on Egyptian tombs and monuments. In 1899, Carter was appointed to the position of Chief Inspector of the Egyptian Antiquities Service. He then began a series of scientific investigations that were ultimately, under the sponsorship of Lord Carnarvon, to lead him to the tomb of Tut-ankh-Amun, who ruled Egypt for a brief period more than three-thousand-two-hundred years ago. The splendour of that tomb, along with its extraordinary furnishings—five-thousand different objects within it were discovered by Carter—revealed a golden age of arts and crafts unequal to any other period in antiquity.

When Carter entered the tomb in the Valley of the Kings, in November 1922, it led him and countless others in due course to recreate an interest in the glories of Ancient Egypt. As *The New York Times*[5] reported on the day of Carter's death in 1939, his discovery, sensational as it was, was more than an archaeological find: *'It was the revealing in dramatic miniature of the whole civilization of an ancient race in a manner that brought out the human rather than the scientific aspects.'*

At his RI Discourse in 1925, Carter explained how he and his team had cleared an estimated seventy-thousand tons of sand and gravel in what seemed a futile search for a new tomb. But, at the corner of the excavated tomb of Ramesses VI,

Figure 9.4 *Howard Carter, the most renowned archaeologist ever, who lectured on the tomb of Tut-ankh-Amun at the RI in 1925. It is a measure of the great attractive power of the house in Albemarle Street that one of the celebrities in greatest demand at that time could be persuaded to present a Discourse there.*
(Wikipedia open access)

a workman found a step cut in bedrock. Subsequent work by Carter and his associates led to a steep flight of stairs leading downward to the tomb.

Carter showed forty-four slides to illustrate the conditions and magnificence of the burial chamber. He also described how almost insufferably hot it was and how congested was the space containing the objects within the shrine. The tomb consisted of four chambers, each enclosed by golden doors and containing more than six-hundred groups of precious objects. In the innermost chamber was the sarcophagus, containing the mummified body of the king. Within the monument, the king lay enclosed in three coffins, rested one within the other, and each forming the (now well-known) effigy of the monarch.

Tut-ankh-Amun's body[6] was bound in bandages of fine linen, with his head protected by the solid gold mask, now exhibited in the Cairo Museum, and familiar to most citizens of the world.

Two of the slides shown by Carter at his Discourse are of special interest: one for its scientific uniqueness, the other for its artistic elegance—see Figures 9.5 and 9.6. In Figure 9.5 is shown one of the two daggers that were placed on the king's abdomen in his sarcophagus. This dagger (Figure 9.5(a)) is made of meteoric iron. It contains a substantial quantity of nickel, a fact that signifies that its origin is extraterrestrial. It is interesting to note that the literal translation of the hieroglyph for iron is *'the metal from heaven'*. Nickel forms an alloy with the iron that is far less likely to corrode than either of the two components, iron and nickel

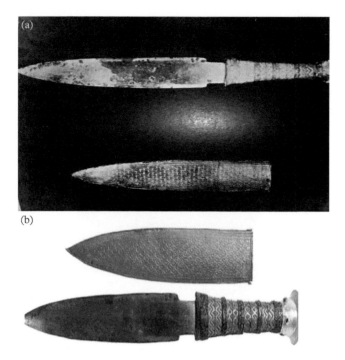

Figure 9.5 *(a) Dagger made of meteoric iron found in the tomb of Tut-ankh-Amun, with sheath. (b) Colour reproduction of the dagger that has undergone little corrosion, and its accompanying sheath.*

((a) Black-and-white reproduction from Griffith Institute, Oxford; (b) provided by Professor Zaki Iskander)

(rather like stainless steel). This may have been why the meteoric iron dagger was included in the tomb; because it resisted corrosion, like the gold that was there in relative abundance.

9.3.1   The Chalice-Like Cup in Tut-Ankh-Amun's Tomb

The article written in the *Proceedings of the Royal Institution*[7] contains the following passage:

'This great (outermost) shrine was overlaid with finely incised gold foil and inlaid with brilliant blue faience tiles decorated with protective symbols djed and tyet. The level of the floor of the chamber was about three to four feet lower than that of the Ante-Room. Around this great shrine, resting on the ground, there were a number of funerary emblems and beautiful objects. Among them the more important were: In the south-east corner of the chamber, Amun's Sacred Goose (Chenalopex Aegyptiacus), made of wood and varnished with a black resin; beside it stood a lamp carved out of pure semi-translucent calcite. This lamp of chalice form, flanked the fretwork ornament symbolising "unity" and

Figure 9.6 *(a) Black-and-white photographs of the calcite chalice where (left) the wick in the oil is not lit and (right) where it is. (b) B Colour reproduction of the calcite chalice shown by Howard Carter at the Discourse at the RI in 1925. (It was also shown by the present author[8] in his Discourse at the RI in 1976.)*

(Courtesy Cairo Museum; photograph Sandro Varnini)

"eternity", ranks among the most interesting and unique objects discovered in the Tomb. It's chalice-like cup, which held the oil and floating wick, was neither decorated on its exterior nor interior surfaces, yet when the lamp was lit you saw the king and queen in brilliant colours within the thickness of the semi-translucent calcite. The explanation would seem to be that there were two cups turned and fitted one within the other, and on the outer wall of the inner cup a picture had been painted in semi-translucent colours, seen only when the lamp was lit."

9.4   Zaki Iskander and Cyril Aldred

Professor Zaki Iskander was the Director of Antiquities at the Cairo Museum and a Professor of the American University when I first met him in Cairo in 1973. He was originally a chemist by training, and he completed his Ph.D. under the supervision of Professor C. K. Ingold, studying the phenomenon of three-carbon tautomerism, before turning his attention to Egyptology.

When I was invited by George Porter to present a Discourse at the RI, I wrote to Professor Iskander, as I wished to show the dagger (of iron) contained in the tomb of Tut-ankh-Amun. Figure 9.7 is an excerpt from the letter written to me (at Aberystwyth, where I then worked) by Professor Iskander. This contains the symbol (hieroglyph) for iron, and Professor Iskander's explanation of its literal meaning.

Cyril Aldred became one of the world's leading Egyptologists. He graduated in English from King's College, London, then studied art history at the Courtauld Institute. But in due course, he flourished as a world-leading Egyptologist, and was Curator at the Royal Scottish Museum, in Edinburgh, where he worked for the remainder of his professional life, apart from spending a year at the Metropolitan Museum of Art, in New York. His book *'Akhenaton, Pharaoh of Egypt'*, published in 1968, became a classic, and his *'Jewels of the Pharaohs'*, published in 1971, is in my opinion one of the most fascinating accounts of the science, art, culture, and the everyday lives of tomb-robbers in Ancient Egypt. Thanks to Champollion, we are able to access records of criminal proceedings conducted in Thebes (present-day Luxor) more than three-thousand years ago.

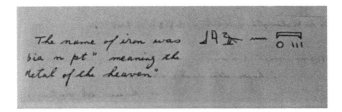

Figure 9.7 *Except of letter from Zaki Iskander to the author (1976), showing the hieroglyph for iron. (Private communication)*

Included in Aldred's book '*Jewels of the Pharaohs*' is the actual testimony of an ancient Egyptian tomb-robber, Amun Pnufer, under rigorous examination in the time of Ramasses IX (*ca.* 1124 BC): '*We found the noble mummy of the sacred king…golden ornaments were upon his breast and a golden mask was over his face…we collected the gold together with the amulets and jewels that were about him…then we set fire to the coffins.*'

Aldred's text is accompanied by a remarkable group of fantastic jewellery that escaped the tomb-robbers' clutches.[9]

9.5   Envoi

Howard Carter's name will live forever in the annals of archaeology, and among all students of antiquity. There are YouTube videos available of him, which show him to be a charismatic individual with a gift of lucid exposition. Strangely, for one so famous, he never received a civil honour from his own country, a distinction he shared with writer Charles Darwin and botanist and anthropologist Thomas Henry Huxley. It is also rather sad to read[4] in T. G. H. James' definitive book published in 1992 that Carter, arguably the most renowned archaeologist of all time, was a tragic human being, unhappy partly because of his latent homosexuality.

APPENDIX 1

Piazzi Smyth and the Pyramid Inch[10]

Piazzi Smyth, motivated by his own religious views, set out for Egypt to test his belief that the Great Pyramid of Khufu was built with just enough 'pyramid inches' to make it a scale model of the circumference of the Earth, and that the perimeter measurement corresponded exactly to the number of days in the solar year. These ideas were tied to his belief that the British inch was derived from the ancient pyramid inch, and that the cubit used to build both Noah's Ark and the Tabernacle of Moses was also based on this inch. These views were firmly discredited by numerous bodies including the Royal Society, which prompted Piazzi Smyth to resign from being a Fellow.[9]

APPENDIX 2

A Summary of How Ancient Egyptians Used Coloured Minerals

In Section 9.2 above, reference was made to the ways in which the Ancient Egyptians selected the minerals that they included in their exquisitely beautiful jewellery, such as the scarab bracelets that were in the tomb of Tut-ankh-Amun (from Thebes), shown in Figure 9.8.

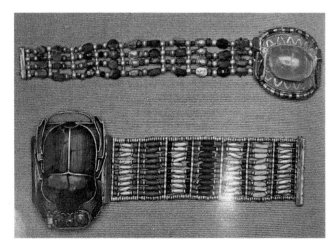

Figure 9.8 *Scarab bracelets, like these shown here, were found in the tomb of Tut-ankh-Amun. Reproduced from Cyril Aldred's book 'Jewels of the Pharaohs',*[9] *Thames and Hudson, 1971. (Book reproduction Thames and Hudson, awaiting permission)*

When I gave a Discourse at the RI in 1976 on *'Adventures in the Mineral Kingdom'*,[8] I showed this picture, as well as one containing scarab and falcon pendants, also from the tomb of Tut-ankh-Amun.

The Ancient Egyptians made extensive use of minerals in a variety of contexts, to beautify their palaces and other belongings, as well as the walls and contents of their tombs.

The colour of naturally occurring minerals, and their synthetic equivalents, was far more important to them than their other properties, such as brilliance or rarity. Carnelion, as mentioned earlier, with its red colour was thought by the Egyptians to preserve within it the colour of lifeblood. Carnelion is silica, SiO_2, which contains traces of iron oxide that it confers upon its colour. The green of beryl (or of malachite) represented the fresh green of springing vegetation. And the blue of azurite or lapis lazuli (or their artificial non-crystalline Egyptian faience) signified the blue of water and the holy sky-realm.

The Ancient Egyptians are believed to be the first to produce a synthetic pigment, namely Egyptian blue, which was first used in pre-dynastic days, more than five-thousand years ago. It was used extensively for decorative purposes, and when the Romans conquered Egypt they used it, making it by the formula devised by the Egyptians:

$$Cu_2CO_3(OH)_2 + 8SiO_2 + 2CaCO_3 \rightarrow CaCuSi_4O_{10} + 3CO_2 + H_2O.$$

They did this by heating malachite ($Cu_2CO_3(OH)_2$ in the presence of silica (sand) and the flux known as quicklime, CaO, generated when calcium carbonate is heated to high temperature.

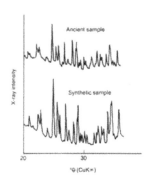

X-ray diffractograms showing the similarity in pattern between Egyptian blue synthesised in the laboratory and that used by the ancient Egyptians for the head in the bust

Figure 9.9 *Sir Humphry Davy rediscovered the recipe for the mineral-based pigment known as Egyptian blue, which was used in the fourteenth century BC to decorate Nefertiti's crown (left). The X-ray fingerprints (right) signify that Egyptian blue is calcium copper silicate.*[12]
(Courtesy American Philosophical Society)

Malachite, which occurs extensively in the Sinai, was mined by many ancient civilizations—in Russia, Mexico, and in the UK. (There was a malachite mine in the Great Orme, Llandudno, in Neolithic times, three-thousand-eight-hundred years ago.)

The secret to the synthesis of Egyptian blue became lost shortly after the Roman Empire collapsed. However, Humphry Davy, in 1815, worked out afresh[11, 12, 13] how to produce it (see Figure 9.9).

It is interesting to note the title of Davy's paper, '*Some Experiments and Observations on the Colours Used in Paintings by the Ancients*', as Nerfertiti's head shows Egyptian blue in all its glory.

Those who have had the privilege of entering the tomb of Nefertari, Ramasses II's favourite wife, in the Valley of the Queens, are also overwhelmed by the sheer brilliance of the mineral pigments used in that tomb, which dates from the thirteenth century BC.

REFERENCES

1. Alternative names for of Khufu, Khafre, and Menkaure are Cheops, Chephren, and Mikerinos.
2. This was part of social entertainment he often gave at evening parties in London at that time.
3. His contribution to astronomy was quite major. He was the first to advocate so-called 'mountain astronomy'. His experiments in 1856, about which he spoke at an RI Discourse

in March 1857, on the *'Peak of Tenerife'*, demonstrated that there—to borrow Isaac Newton's words—in *'serene air above the grosser clouds'* lay the future for observational astronomy. This is the reason why modern optical astronomers place their telescopes at high altitudes in various parts of the world.

4. T. G. H. James, *'Howard Carter and the Path to Tut-ankh-Amun'*, Kegon Paul, Ichdon and the American University in Cairo Press, Cairo, **1992**.

5. 'Howard Carter, 64, Egyptologist Dies', *The New York Times*, 3 March, 1939.

6. The archaeologist Mary Crowfoot examined the linen that was found in the tomb of Tut-ankh-Amun. She was the mother of Dorothy (Crowfoot) Hodgkin (born in Cairo), who visited the RI as a young scientist, and who later gave a Discourse there on her determination of the structure of vitamin B_{12}, one of the achievements that earned her the Nobel Prize in Chemistry in 1964.

7. H. Carter, *'The Tomb of Tut-ankh-Amun'*, Proc. Roy. Inst. Great Brit., **1925**, 636–645.

8. J. M. Thomas, *'Adventures in the Mineral Kingdom'*, Proc. Roy. Inst. Great Brit., **1976**, *49*, 243.

9. C. Aldred, *'Jewels of the Pharaohs: Egyptian Jewllery of the Dynastic Period'*, Thames and Hudson, London, **1971**.

10. M. Lehner, *'The Complete Pyramids'*, American University in Cairo Press, Cairo, **1997**, p. 57.

11. H. Davy, *Phil. Trans. R. Soc.*, **1815**, *105*, 97.

12. J. M. Thomas, *Proc. American Phil. Soc.*, **2006**, *150*, 528.

13. J. Ragai, *'Substantia'*, Florence University Press, **2018**, p. 93.

10

Peter Mark Roget: Facilitator of the Writing of Good English and Two of his Distinguished Successors as the Fullerian Professor of Physiology (Thomas Henry Huxley and Sir Peter Medawar)

10.1 Introduction

Close to the time of my beginning to write my Ph.D. thesis in 1957, my research supervisor told me to purchase a copy of '*Roget's Thesaurus*'. He assured me that it would assist me in making my thesis more readable. Until that time, I had never heard of the word 'thesaurus', and I knew not who Roget was. On purchasing a copy of the abridged version, I quickly warmed towards it. First, there was the sub-heading 'claim' on the title page: '*Classified and Arranged so as to Facilitate the Expression of Ideas and to Assist in Literary Composition*'. In the Preface, Roget had written that the function of a thesaurus was the converse of that of a dictionary. A thesaurus starts from the 'idea'; it then '*finds the word or words by which that idea may be most fitly and aptly expressed*'. This was a great help for me. Like countless other thesis writers, I found '*Roget's Thesaurus*' indispensable.

It was some twenty years later, when I was first taken to attend a Friday Evening Discourse by Sir Gordon Cox,[1] that I heard of Roget again. It was explained to me that when Michael Faraday became the first Fullerian Professor of Chemistry at the Royal Institution (RI), in 1833, another Fullerian Professorship was also established, in physiology; and the first occupant of that post—a permanent one, but unlike Faraday's, non-resident—was Peter Mark Roget.

These Fullerian Professorships were created at the RI as a result of a generous bequest by an eccentric English Member of Parliament (MP), John Fuller (see

Albemarle Street: Portraits, Personalities, and Presentations at the Royal Institution. John Meurig Thomas, Oxford University Press. © Sir John Meurig Thomas 2021.
DOI: 10.1093/oso/9780192898005.003.0010

Figure 10.1 *John Fuller, MP, founder of the Fullerian Professorship of Chemistry and Physiology at the Royal Institution (RI).*
(Courtesy the RI)

Figure 10.1). It was said of Fuller '*...the feebleness of whose constitution denied him at all other times and places the rest necessary for health could always find repose and even quiet slumber amid the murmuring lectures of the Royal Institution and that, in gratitude for the peaceful hours thus snatched from an otherwise restless life, he bequeathed to the Royal Institution the magnificent sum of £10,000.*'

10.2   The Unusual Professor Roget

Roget was born in Lincoln in 1779, the son of a Genevan cleric, but he moved to Edinburgh in 1793, where he later studied medicine at the university. Shortly, thereafter, he moved south, and while he was resident in Clifton, near Bristol, he became acquainted with both Thomas Beddoes of the Pneumatic Institute, (see Chapter 3) and with Humphry Davy, whom Beddoes had hired. Later, after a visit to continental Europe, Roget became a private physician to the First Marquis of Lansdowne, after whose death in 1805 he was appointed a lecturer in physiology at Manchester Infirmary. In 1808, he moved to London and became a licentiate of the Royal College of Physicians. In due course, he began to give public lectures, some of which were delivered at the RI.

In 1828, along with Thomas Telford, the civil engineer, and William Thomas Brande, who had been appointed to the RI to replace Humphry Davy after the latter's formal retirement in 1812, Roget submitted an important report on London's water supply, a topic that later engaged the interest of Michael Faraday. Five years later, he and Faraday were appointed to their respective Fullerian Professorships at the RI.

On the basis of an impressive paper that he had written on a slide rule with a log-log scale that he had devised, Roget was elected a Fellow of the Royal Society in 1815. From 1827–48, he served as Secretary to the Royal Society, which meant he overlapped briefly with Davy and with the other Cornishman, Davies Gilbert, as President.

Roget's other scientific interests involved questions related to persistence of vision. He invented the so-called phantamascope, an instrument that could trace, stroboscopically, animation in various living creatures.

Roget was a public-spirited individual who was involved in the establishment of the Society for the Diffusion of Useful Knowledge,[2] and he promoted the work of the Medical and Chirurgical Society of London.[3] He also contributed an article on natural theology to the *Encyclopaedia Britannica*.

10.3   Roget's Thesaurus

Roget's book, the full title of which is '*Thesaurus of English Words and Phrases*', was first published in 1852, when Roget was seventy-three, and it has remained in print ever since. It is an invaluable collection of synonyms and antonyms and is much respected by most English-speaking people. It is reputed that Roget started giving thought to its creation quite early in his life, well before he retired from his professional career in medicine and physiology. Roget, so it was said, was affected by mental depression and suffered from insomnia, and that the thesaurus arose partly because it suppressed these afflictions. Roget had even begun to maintain a notebook classification scheme for his premier creation as early as 1805. During Roget's lifetime, the work had twenty-eight printings. After his death at ninety, it was revised and expanded by John Lewis Roget, his son, and later by Samuel Romilly Roget, his grandson.

Figure 10.2 *Peter Mark Roget.*
(Courtesy the Royal Society)

10.4   Roget's Successors as Fullerian Professors of Physiology at the RI

Notable holders of the visiting post of Fullerian Professor of Physiology included very many distinguished scientists in British public life. Thus Thomas Henry Huxley, whose work we shall examine a little more fully below, and who was known as Darwin's bulldog—in view of his persuasive and sometimes pugnacious defence of the theory of evolution—was one of them, as were his grandchildren, Julian Huxley, who became an eminent authority on the courtship of birds, and Sir Andrew Huxley, the Nobel Laureate half-brother of Julian, renowned for the Hodgkin—Huxley work on nerve conduction. As well as the Huxleys, the following eminent individuals all held the Fullerian Professorship of Physiology at the RI:

- J. B. S. Haldane, the brilliantly versatile biological scientist and former classical scholar
- Charles Sherrington, the Nobel Prizewinner who coined the word 'synapse'
- William Bateson, who coined the word 'genetics'

- Peter Medawar, the Brazilian-born Nobel Prizewinner who discovered acquired immune tolerance and paved the way to replacement surgery
- Max Perutz, the Austrian-born Nobel Prizewinner for his pioneering studies of haemoglobin and a co-founder of the Laboratory of Molecular Biology of the Medical Research Council in Cambridge
- Anne McLaren, the eminent development biologist whose work helped lead to human in vitro fertilization and winner of the Japan Prize in 2002
- John Gurdon, the Nobel Prizewinning physiologist who pioneered stem cell research
- Francis Crick, of the structure of DNA fame and numerous other contributions to molecular biology
- David Phillips, who led the RI team that determined the structure of the first enzyme (lysozyme)
- Sydney Brenner, for his Nobel Prizewinning work on several areas of molecular biology including the genetic code

Three of these successors to Roget became Presidents of the Royal Society and most of them were awarded its premier medals.

10.5   Some Specific Contributions by Huxley and Medawar

These two giants of physiology were very influential in promulgating their scientific knowledge to lay folk during the course of their visiting professorship. The others mentioned above were also very successful, but Thomas Henry Huxley and Peter Medawar merit special attention.

10.5.1   Thomas Henry Huxley

Born in Ealing, Middlesex, on 4 May 1825, Thomas Henry Huxley began his studies in that town. In 1841, he started studying medical subjects at London University. After gaining his Bachelor of Medicine degree, he joined the Royal Navy and became an assistant surgeon on HMS *Rattlesnake*. This took him to Australia and the Barrier Reef, where he studied the medusa and hydrozoan. His memoirs, presented to the Linnean and Royal Societies, brought him immediate acclaim. Two years later, he was elected Fellow of the Royal Society. Shortly thereafter, his reputation as an anatomist of great authority was established. After the publication of Charles Darwin's '*On the Origin of Species*' in 1859, Huxley developed a reputation of being one of the theory's ardent supporters. Later, as Secretary, then President, of the Royal Society, he was instrumental in organizing the famous *Challenger* expedition,[4] and later served as an Inspector of Fisheries (Figure 10.3).

Figure 10.3 *Portrait of Thomas Henry Huxley, President of the Royal Society and Fullerian Professor of Physiology at the RI.*
(By kind permission of the Royal Society)

Huxley served twice as the Fullerian Professor at the RI, from 1855–8 and from 1865–9. While occupying this post, he became very friendly with John Tyndall, the Irish scientist educated in Germany and Faraday's successor as Fullerian Professor.[5]

Both Huxley and Tyndall had convinced themselves of the veracity of Darwin's theory of evolution, which rests on the proposition of the survival of the fittest. At that time, it seemed to most members of the public that this theory attributed to nature the creative work that had been regarded previously as God's special province. Was nature thus independent of God, and therefore not under His moral law? Huxley and Tyndall, in their public pronouncements, were among the strongest supporters of Darwin's theory.[6]

At the three-hundredth anniversary of the birth of the physician William Harvey[9] in 1878, Huxley began his Discourse at the RI as follows: '*Many opinions have been held respecting the exact nature and value of Harvey's contributions to the elucidation of the fundamental problem of the physiology of the higher animals; from those which deny him any merit at all—indeed, roundly charge him with the demerit of plagiarism— to those which enthrone him in a position of supreme honour among great discoverers of science. Nor has there been less controversy as to the method by which Harvey obtained the results which have made his name famous. I think it is desirable that no obscurity should hang around these questions; and I add my mite to the store of disquisitions on Harvey which this year is likely to bring forth, in the hope that it may help to throw light upon several points about which darkness has accumulated, partly by accident and party by design.*'

Huxley ended on a note which is relevant to the present-day debate between vivisectionists and their opponents!

'The fact is that neither in this, nor in any, physiological problems can mere deductive reasoning from dead structures tell us what part that structure plays when it is a living component of a living body. Physiology attempts to discover the laws of vital activity, and these laws are obviously ascertainable only by observation and experiment upon living things!'

10.5.2   Peter Medawar

The distinguished British biologist and writer Sir Peter Medawar is widely admired, not only by the medical and scientific community for his pioneering work as the father of organ transplantation—through his Nobel Prizewinning work on acquired immune tolerance—but also because of his skill as a writer, expositor, and advocate of the scientific method and related topics. Many scientists from other fields regard Medawar as a superb writer and thinker. Thus, Martin Rees, one of the world's leading cosmologist (the present Astronomer Royal and former President of the Royal Society), has often extolled Medawar's stylish writing and depth of intellectual analysis. Richard Dawkins, himself no mean expositor, regarded Medawar as *'the wittiest of all scientific writers'*. His books are a delight to read, and cover a variety of vitally important topics: *'The Uniqueness of the Individual'*; *'The Future of Man'*; *'Advice to a Young Scientist'*; *'Pluto's Republic'*; and the one I treasure the most *'The Limits of Science'*.

The central purpose of the last-named work is, as he says, to *'exculpate science from the reproach that it is quite unable to answer ultimate questions such as the following: how did everything begin? Where do we come from? What is the point of living? Does God exist?'* Such questions have existed from the days that humankind arrived on Earth. They have prompted numerous answers as well as profoundly moving statements, one of my favourites being from Albert Einstein: *'Strange is our situation here on Earth. Each of us comes for a short visit, not knowing why, yet somehow seeming to a divine purpose. From the standpoint of daily life, however, there is one thing we do know: we are here for the sake of others.'*[7]

Medawar, through his intellectual vitality, perspicacity, and profundity of thought, was a favourite lecturer at the RI in his days there as Fullerian Professor of Physiology.[8] The man who appointed him, George Porter, my predecessor as Director of the RI, told me, as also did regular distinguished members of the RI, such as Max Perutz, David Phillips, and Sir John Gurdon (all former Fullerian Professors of Physiology), that all of Medawar's performances at the RI were exhilarating.

The richness of Medawar's language, as well as the profundity of his philosophical arguments, are well illustrated in the excerpt from his work.

A few further facts about Medawar are relevant to recall. He was the youngest child of a Lebanese father and British mother, and was both a Brazilian and British citizen by birth. He studied in Marlborough College and Magdalen College,

Figure 10.4 (a) *Sir George Porter, OM.*
(Taken from Wikipedia)

Figure 10.4 (b) *Sir Peter Medawar, OM.*
(Taken from Wikipedia)

Oxford, and was Professor of Zoology at the University of Birmingham and University College, London. He became Director of the National Institute of Medical Research at Mill Hill, London, the post he held until he was partially disabled by a cerebral infarction.

Medawar greatly approved of Karl Popper's philosophy. He even went so far, in answer to a BBC interviewer, to state that Popper was among the three greatest living Englishmen, the others being Max Perutz and Ernst Gombrich.[10]

Medawar was a declared atheist. But when he died, a memorial service for him was conducted at Westminster Abbey. To my knowledge, only one non-Christian had been accorded this posthumous honour, namely Charles Darwin. (More recently Professor Stephen Hawking was also given a memorial service in Westminster Abbey.)

At Medawar's memorial service in December 1987, his friend George Porter, the then President of the Royal Society, read from Medawar's *'The Hope of Progress'*:

'We cannot point to a single definitive solution of any one of the problems that confront us—political, economic, social or moral, that is having to do with the conduct of life. We are still beginners, and for that reason may hope to improve. To deride the hope of progress is the ultimate fatuity, the last word in poverty of spirt and meanness of mind. There is no need to be dismayed by the fact that we cannot yet envisage a definitive solution of our problems, a resting-place beyond which we need not try to go. Because he likened life to a race and defined felicity as the state of mind of those in the front of it, Thomas Hobbes[1] has always been thought of as the arch materialist, the first man to uphold go-getting as a creed. But that is a travesty of Hobbes' opinion. He was a go-getter in a sense, but it was the going, not the getting, he extolled. As Hobbes conceived it, the race had no finishing post. The great thing about the race was to be in it, to be a contestant in the attempt to make the world a better place, and it was a spiritual death he had in mind when he said that to forsake the course is to die. "There is no such thing as perpetual tranquillity of mind while we live here," he told us in "Leviathan", *"because life itself is but motion and can never be without desire, or without fear, no more than without sense. There can be no contentment but in proceeding." I agree.'*

The major turning point in his scientific life came when in 1949 he read the work of the Australian biologist Frank Macfarlane Burnet in Melbourne, where Macfarlane advanced the hypotheses that, during embryonic life and immediately after birth, cells gradually acquire the ability to distinguish between their own tissue substances on the one hand and unwanted cells and foreign material on the other.

With a co-worker Billingham, Medawar published a seminal paper in 1951 on grafting techniques. These workers had extracted cells from young mouse embryos and injected them into another mouse of different strains. When this

---

[1] This similie occurs more than once in Hobbes. The passage I have in mind is from his *'Human Nature'* (London, **1650**, Chapter 9).

developed into adult, and skin grafting from that of the original strain was performed, there was no tissue rejection. This meant that the mouse had tolerated
the foreign tissue, which would normally be rejected. This experimental proof of
Burnet's hypothesis was published in 1953, and was followed by a series of
definitive papers and established beyond doubt the reality and subsequent
medical viability of 'actively acquired tolerance', for which Medawar and Burnet
shared their Nobel Prize.

10.5.3   Other Qualities Pertaining to Medawar

Medawar gave the BBC Reith Lectures in 1959, his six talks being entitled *'The
Future of Man'*. He was a man of many interests, notably opera, philosophy, and
cricket. On his views on religion he made the following statement: *'...I believe that
a reasonable case can be made for saying, not that we believe in God because He exists,
but rather that He exists because we believe in Him....'*

Peter Medawar was elected President of the Royal Society in 1970, but he was
unable to take up the post owing to the state of his health.

REFERENCES

1. Sir Gordon Cox, as E. G. Cox, was one of Sir William Henry Bragg's acolytes in the
   1930s at the Davy–Faraday Research Laboratory. Trained as a physicist at Bristol
   University, he left the RI for the University of Birmingham, to work with Sir Normal
   Howarth, whose Nobel Prize was awarded for his determination, by classical organic
   chemical methods, of the structure of vitamin C (ascorbic acid). It was Cox who was
   first to determine the structure of vitamin C by X-ray crystallography. Later, Cox
   became Professor of Inorganic Chemistry at the University of Leeds, where he established a thriving School of Crystallography. Later still, he became Secretary of the
   Agriculture Research Council. He was Treasurer of the RI when I took over from
   George Porter as Director in 1986.
2. The Society for the Diffusion of Useful Knowledge was founded in 1826 in London
   with the object of publishing information to people who were unable to obtain formal
   teaching or who preferred self-education.
3. This society was founded in London in 1805 by twenty-six medics who had left the
   Medical Society of London. In addition to Roget, the society's doctor, Alexander Marcet,
   husband of Jane Haldimand Marcet, the author of *'Conversations in Chemistry'* (see
   Chapter 4), was also a founder member.
4. The *Challenger* expedition of 1872–6 was a scientific programme that made many discoveries to lay the foundation of oceanography. The mother vessel was HMS
   Challenger. Among many other discoveries, this expedition catalogued more than
   four-thousand previously unknown species.
5. R. Jackson, *'The Ascent of John Tyndall'*, Oxford University Press, **2019**.
6. By John Murray in Albemarle Street.
7. I am grateful to Professor Dudley Herschbach for drawing this statement to my attention.

8. Medawar preferred that his visiting post at the RI was designated Professor of Experimental Medicine.

9. William Harvey (1578–1659), English anatomist and physiologist who was first to discern blood circulation and the heart's function as a pump. His *'De Motu Cordis'* (1628) is regarded as the foundation of modern medicine.

10. All three having been born in Vienna.

11

The Most Beautiful Experiment in Physics: Candidates from the RI and Elsewhere

11.1 Introduction

It is often the case that university and other teachers raise the question as to the most impressive, or most beautiful, or most intellectually transformative, experiment that they have encountered. In my own experience, I find that discussions with other educators and disseminators of scientific knowledge among lay audiences lead in turn, to lecture-demonstrations as the ones that drive home their message most effectively.

I contend that four individuals who worked in the laboratories of the Royal Institution (RI) in Albemarle Street are candidates that merit serious consideration in this context: Thomas Young, Humphry Davy, Michael Faraday, and Lawrence Bragg. Before I proceed to consider each of these candidates, it is prudent to mention other experiments, some of ancient lineage, which are often raised in discussion pertaining to this fascinating topic.

In this connection, the reader may wish to consult an interesting relevant article by R. P Crease published in *Physics World* in 2002.[1] That article takes the long view and enumerates some of the classics in physics that involve Galileo, Newton, Foucault, Rutherford, Millikan, and Joule. One of the experiments—a favourite of mine—was carried out in pre-Christian times by Eratosthenes, the head librarian of the Great Library of Alexandria, and I consider it worthy of summarizing later in this chapter.

11.2 Thomas Young: Phenomenal Young

Before we describe this experiment that was carried out at the RI in 1801, it is appropriate first to describe Young and his astonishing qualities. As mentioned

Albemarle Street: Portraits, Personalities, and Presentations at the Royal Institution. John Meurig Thomas,
Oxford University Press. © Sir John Meurig Thomas 2021.
DOI: 10.1093/oso/9780192898005.003.0011

earlier, he was born in Milverton, Somerset, in 1773 and was brought to the RI by its discriminating founder, Rumford, as Professor of Natural Philosophy.

Young was a precocious child who taught himself natural history, natural philosophy, fluxional calculus, and how to assemble microscopes and telescopes. He subsequently learnt Hebrew, Chaldean, Syriac, Samaritan, Arabic, Persian, Turkish, and Ethiopic (but, unlike Champollion, another remarkable linguist, not Coptic, which, as we saw in Chapter 9 enabled Champollion to proceed further than Young in deciphering the Rosetta Stone).[2] Young read medicine in London, Edinburgh, and Gottingen, and then he moved to Emmanuel College, Cambridge, for two years, where he was known as 'Phenomenon Young'. He was not a success as a general practitioner (GP) prior to his recruitment by Rumford to the RI, and his lectures there were far less popular than those of his contemporary Davy.

At the RI, Young made many vital experimental contributions to such topics as capillarity, the spreading of liquids on solid surfaces, and the significance of the contact angle of an oil drop on a solid surface. He made important studies also in optics and vision, which included the discovery of astigmatism and, more significantly, proposed the three principal colours that rationalize colour vision. He also discovered how the ciliary muscles of the eye aided vision. In addition, in his *Encyclopaedia Britannica* article on languages, Young compared the grammar and vocabulary of some four-hundred tongues! His article on Egypt in the 1818 edition of *Encyclopaedia Britannica* disclosed that he had identified the hitherto unidentifiable cartouche on the Rosetta Stone as the symbol for the Pharaoh Ptolemy V, who issued the trilingual stone—Hieroglyphic, Greek, and Demotic—as a decree from Memphis in 196 BC. It has often been said that Thomas Young was the last man who knew everything. The omnivorous Lord Rayleigh (Chapter 5) revered him. See Table 11.1.

11.2.1 The Double-Slit Experiment

This experiment was carried out by Young to demonstrate the wave nature of light. Waves give rise to the phenomenon of interference. So, in order to show that Newton's view, which prevailed at that time, was erroneous—it was based on the assumption, as expressed in Newton's '*Opticks*', that light was particulate—Young devised the double-slit experiment. He first of all showed the reality of interference of waves by using a ripple tank, in which two sources of waves on water were used. Then he used two sources of light (as shown in Figures 11.1 and 11.2).

What is rather remarkable about Young's double-slit experiment is that it can now be readily carried out not with light but with a stream of single electrons. Remarkably, interference patterns (just as with light) are generated from a primary beam of single electrons. Modern physicists interpret this result as the demonstration of wave-particle duality, as well as quantum physics itself. It demonstrates that single electrons proceeding one-by-one interfere. Richard Feynman is said to

Table 11.1 *Topics on which Thomas Young Offered to Contribute Articles to* Encyclopaedia Britannica

Alphabet	Friction
Annuities	Haloes
Attraction	Hieroglyphics
Capillary Action	Hydraulics
Cohesion	Motion
Colour	Resistance
Dew	Ships
Egypt	Sound
The eye	Strength
Focus	Tides
	Waves

And anything of a medical nature.

ᵃ (Compiled from A. Robinson, *Nature*, **2005**, *438*, 291.)

Figure 11.1 *Thomas Young.*
(Courtesy the Royal Society)

have remarked that it contains everything you need to know about quantum mechanics.

Young's double-slit experiment played a crucial role in the work of Ahmed Zewail, the Nobel Prizewinning founder of the field now known as femtochemistry, which deals with the science of events that take place in femtoseconds (fs) (1 fs ≡ 10^{-15} s).

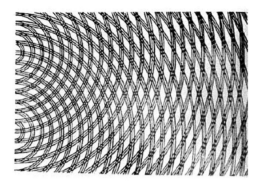

Figure 11.2 *Young's drawing to illustrate the interference of waves from two light sources.*
(Courtesy Ahmed H. Zewail)

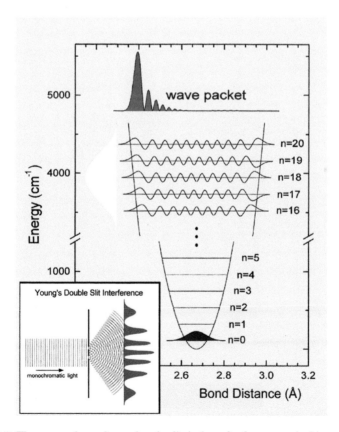

Figure 11.3 *The wave packet and wave function limits for molecular systems, in this case a diatomic molecule.*
(Courtesy Ahmed Zewail)

(A nanosecond (ns) is 10^{-9}s, while a femtosecond is a million times as fast as a nanosecond. To appreciate how fast this is, recall that there are as many nanoseconds in a second as there are seconds in a human being's life.) Zewail, the Linus Pauling Professor at the California Institute of Technology, also gave a brilliant Discourse at the RI (see Figures 11.2 and 11.3).

Note that Figure 11.2, taken from Zewail's Nobel Lecture, contains a drawing of Young's experiment. It is only by using Young's double-slit principle that Zewail (and later co-workers) can now measure transition-complexes, which form

Figure 11.4 *Dr Ahmed Zewail receiving the Nobel Prize in Chemistry from King Carl Gustaf XVI in 1999.*

(Photograph used with permission from the Zewail family)

whenever two or more molecular reactants proceed to give different products. In timescales that fall in the range of 10^{-15} to 10^{-13} seconds, transition states undergo changes in positions of the constituent atoms involved in the chemical change. Zewail was the first to show how, with ultra-fast lasers, and a sophisticated application of Young's experiment, he was able to track the nature and changes of transition states.

11.3   Faraday and Davy

Some of my friends regard Michael Faraday's demonstrating of lines of force surrounding a magnet by sprinkling iron files on it (see Figure 4.6 of Chapter 4) as a candidate. Others propose his proof of electromagnetic induction and also the Faraday effect, in which the plane of polarization of light is altered by the presence of a magnetic field.

I believe that Humphry Davy's lecture-demonstration, in which he wished to illustrate that the electrical resistance of a wire increases with increasing temperature, is an extremely impressive one. He starts by having a metal wire carrying an electric current that is just large enough to make the wire glow a barely visible (in a darkened room) dull red. He then takes a lump of ice and vigorously strokes one of the end regions of the wire with it. Immediately, the other end of the wire glows brightly, just short of incandescence. Reason: lowering the temperature of the wire with ice decreases its electrical resistance, hence the current (with the same applied potential) increases markedly, and causes the wire to glow brightly.[3]

Humphry Davy's considerable work also suggests other candidates. In Chapter 3 (Figure 3.6), we showed the account that Davy gave of his isolation and discovery of elemental potassium, an achievement that he made by using electricity to electrolyse potash—the reverse in effect of what Volta had reported in 1800—as described in his letter to Sir Joseph Banks. Davy went on to identify and discover several other alkali metals and the alkaline earths, as mentioned also in Chapter 3.

Both Mendeleev and Berzelius praised the work in ultra-generous terms: Mendeleev, sixty years later than Davy's work, went so far as to call it *'one of the greatest advances in the whole of science'*. One must concede that what Davy did, and reported in his series of Bakerian Lectures to the Royal Society from 1807 onward, led to the recognition that a whole group of new elements—the top left-hand corner—of the periodic table owes everything to him.

Davy, as described also in Chapter 3, was responsible for carrying out other experiments that are also contenders for the most memorable, specifically those involved in: (i) his invention of the miner's safety lamp; (ii) his associated invention giving light in explosive mixtures of firedamp in coal mines; (iii) his invention of cathodic protection; and (iv) the discovery of the carbon arc; all are also very memorable achievements.

 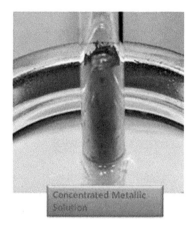

Dilute Electrolytic Solution

Concentrated Metallic Solution

Figure 11.5 *The image of the blue colour shown here on the left is taken from a paper by Professor Peter P. Edwards. It represents the blue colour shown to Oliver Sacks by his Uncle Dave; and it arises from an electron solvated by liquid ammonia.*
(Courtesy Professor Peter P. Edwards)

But others regard another of Davy's experiments—his demonstration of the absorption/solution of alkali metals in liquid ammonia. This is how Oliver Sacks (who repeated the experiment as a boy) describes this remarkable experiment, which was brought to his attention by his Uncle Dave, *'Uncle Tungsten'*.

'The most mysterious and beautiful of all the blues for me was that produced by dissolving alkali metals in liquid ammonia. The fact that metals can be dissolved at all was startling at first, but the alkali metals were all soluble in liquid ammonia (some to an outstanding degree—caesium would completely dissolve in a third of its weight of ammonia). When the solutions became more concentrated they suddenly changed character, turning into lustrous bronze-coloured liquids that floated on the blue—and in this state they conducted electricity as well as a liquid metal like mercury. The alkaline earth metals would work as well, and it did not matter whether the solute was sodium or potassium, calcium or barium—the ammoniacal solution, in every case, became an identical deep blue, suggesting the presence of some substance, some structure, something common to them all. It was like the colour of the azurite in the Geological Museum, the very colour of heaven.'

We now know, thanks to the work of Professor Peter P. Edwards at Oxford, that the *'something common to them all'* is that the colours arise from a solvated electron (an electron in a cavity circumscribed by four molecules of ammonia).

11.4 Lawrence Bragg and Max von Laue

Another candidate involves, first, Max von Laue and his two colleagues in Munich in 1912. This entails the phenomenon of X-ray diffraction. It was von Laue who, with the assistance of a bright graduate student, P. P. Ewald, deliberately set out to ascertain whether X-rays were waves or particles. The fact that a spot pattern (like those shown in Figure 11.4) was obtained (with a crystal of zinc blende, ZnS, and the gemstone, beryl) proved that X-rays could be thought of as waves[4,5].

However, it was the brilliance of the twenty-two-year-old Lawrence Bragg that enabled[6] the simple equation (nλ = 2d sin θ) to be derived (see Figure 11.6).

By regarding layers of atoms in all solids as, effectively, mirrors that reflected X-rays, Bragg opened up a vast new field that had been initiated by von Laue. As described elegantly by R. P. Crease,[7] this experiment led, in turn, to the determination of the wavelengths of X-rays (from different sources). It also

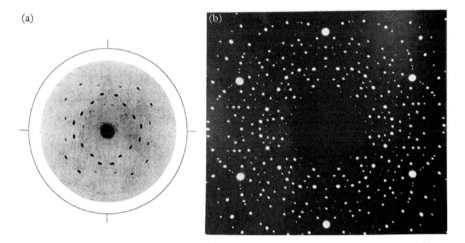

Figure 11.6 *Diffraction pattern produced when X-rays strike a crystal (a) of zinc blende (after von Laue, 1912), and (b) when it strikes the gemstone beryl.*
(Copyright J. M. Thomas)

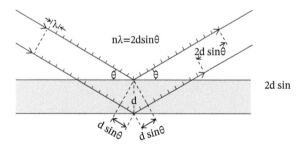

Figure 11.7 *(a) Lawrence Bragg's interpretation of how diffraction arises from reflection at atomic planes, and a statement of Bragg's Law (copyright Joan Starwood). (b) Illustration of the concept behind Bragg's equation.*
(Copyright J. M Thomas)

led to X-ray spectroscopy that Henry Moseley and Manne Siegbahn subsequently pursued. This X-ray spectroscopic work led to the realization of the importance of atomic numbers, which transformed our understanding of the nature of matter, and rationalized further the significance of Mendeleev's periodic table.

Moreover, the pioneering work of both Lawrence Bragg and his father, Sir William Henry Bragg, as well as of all their colleagues at the RI—a world centre for X-ray diffraction for several decades—has led to the award of dozens of Nobel Prizes, earned through the imaginative application of Bragg's Law $(n\lambda = 2d \sin \theta)$ [8,9].

11.5   Eratosthenes and his Measurement of the Circumference of the Earth, Third Century BC

Eratosthenes was born in 276 BC in Cyrene, now part of Libya. He went to Athens to further his studies, and it quickly became apparent that he was a polymath. Later in life, he was reckoned an authority on mathematics, astronomy, poetry, music, and geography.[10] He became the head librarian of the Great Library of

Figure 11.8 *Image of Eratosthenes.*
(Wikipedia open access)

Alexandria. Nowadays, he is renowned as the individual who made a remarkably accurate estimate of the circumference of the Earth. (He is also known for a crater of the Moon named after him, and for the so-called sieve of Eratosthenes, which deals with a method of determining prime numbers.)

Around 240 BC, he made his famous experiment to estimate the circumference of the Earth. It was accomplished by making presumably simultaneous measurements of the angles of the shadows cast by a vertical stick at Syene (present-day Assan) and another at Alexandria at noon on the day of the summer solstice. From the measurements, and knowing the distance from Syene to Alexandria along the assumed same meridian of longitude, Eratosthenes was able to derive a quite accurate estimate of the radius (hence circumference) of the Earth. Figure 11.9 shows the essential features of the information that can be used and the way he arrived at the magnitude of the circumference. Several experts, e.g., Fred Hoyle[11] and others,[12,13] have commented favourably on Eratosthenes' ingenuity.

It is relevant to mention[12] that Eratosthenes decided upon his method when he learned about a most unusual occurrence that took place each year in a well on

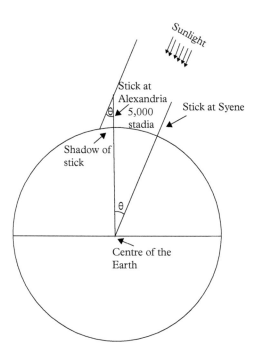

Figure 11.9 *Eratosthenes' measurement of the circumference of the Earth as viewed when the shadow of the stick at Alexandria is at a minimum at noon on the day of the summer solstice at Syene. (Copyright,* The Physics Teacher, *49 (7), 445–447.)*

(R.A. Brown and A Kumar (2011), AIP Publishers permission request)

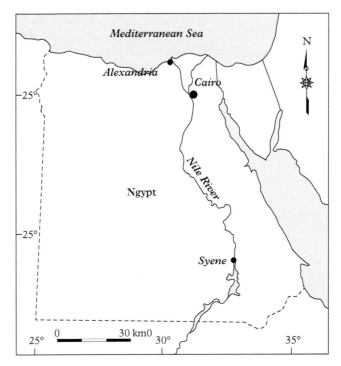

Figure 11.10 *A map of Egypt. The ancient city of Syene is today known as Aswan. (Copyright,* The Physics Teacher, 49 (7), 445–447.)

(R. A. Brown and A Kumar (2011), AIP Publishers permission request)

the island of Elephantine (in Aswan) on the River Nile: '*At noon, on that island, on the day of the summer solstice the sun was said to light up the well right down to the water and cast no shadow on the side.*' Therefore, argued Eratosthenes, the Sun at the place of the well was directly overhead and that the well was located exactly, or very close to, the Tropic of Cancer. This then enabled him to proceed, as indicated diagrammatically in Figure 11.9.

The definition of the value of the 'stade' (plural 'stadia') as a unit of distance in antiquity has been discussed by others. And Hoyle[11] provided an argument for taking Eratosthenes' stade as about 157 m. Since the distance from Syene to Alexandria was known to be approximately 5,000 stadia, the circumference of the Earth could then be calculated as 50 × 5,000 = 250,000 stadia (Eratosthenes had measured the angle θ to the 7.2°, or one-fiftieth part of a circle).

The original paper by Eratosthenes has been lost. Fortunately, isolated portions of it were quoted or rephrased by later, ancient writers and historians such as Cleomedes, Strabo, and Pliny.[14]

REFERENCES

1. R. P. Crease, *Physics World,* **2002**, September, 19.
2. Great rivalry developed between Young and Champollion in the quest to decipher the Rosetta Stone. The Frenchman won the race because he knew the Coptic language.
3. In Lecture 3 of my 1987 Christmas RI Lectures, I repeated Davy's elegant demonstration. It can be viewed on the video of my series (with Professor David Phillips) available on the RI Channel.
4. When Wilhelm Röntgen discovered X-rays in 1895 in Wurzburg, it was as a result of an accident; it is a classic example of a chance discovery. In the case of X-ray diffraction, there was no question of an accident; von Laue argued that if X-rays were waves, they ought to exhibit interference (just as Young had argued concerning the nature of light). It was Paul Ewald who convinced von Laue that the spaces separating atoms in a crystal were likely to be of the same magnitude as the wavelength of X-rays. This is what prompted him to carry out his famous experiment, reported in ref [5].
5. W. Friedrich, P. Knipping, and M. Laue, *Bayerische Akademie der Missenschaften,* **1912**, 303.
6. W. L. Bragg, *Proc. Cam. Phil. Soc.,* **1912**, *17,* 43.
7. R. P. Crease, *Physics World,* **2012**, September, 2.
8. I. Olovsson, A. Liljas, and S. Lidin, '*From a Grain of Salt to the Ribosome: The History of Crystallography as Seen Through the Lens of the Nobel Prizes*', World Scientific Publishers, Singapore, **2015**, p. 3.
9. J. M. Thomas, '*Architects of Structural Biology: Bragg, Perutz, Kendrew and Hodgkin*', Oxford University Press, **2020**.
10. Many also regard Eratosthenes as the father of geography.
11. F. Hoyle, '*Astronomy*', Crescent Books, New York, **1962**, p. 84.
12. R. A. Brown and A. Kumar, *Phys. Teacher,* **2011**, *49,* 445.
13. Diller, *Isis,* **1949**, *40,* 6.
14. J. Dulka, *Anch. Hist. Exact Sci.,* **1993**, *46,* 55.

12

The Uniqueness of the RI: Some autobiographical reminiscences of my days as Director of the RI

12.1 Introduction

This chapter briefly discusses George Porter as occupant of the posts of Director and Fullerian Professor of Chemistry at 21 Albemarle Street. It alludes to his excellence as a scientist and as a Director, who was able to use a small laboratory in a converted house to function as a world-class centre of research in photochemistry and photophysics for a period of thirty years. It is also a personal selection of portraits of the personalities of several key individuals whom I interacted with during the twenty years that I spent in Albemarle Street, 1986–2006. Some of the Discourses presented in this period are also highlighted, as are many other aspects of the affairs of the Royal Institution (RI) whilst I was Director.

12.2 George Porter

One of the greatest physical chemists of the twentieth century, a Nobel Laureate for his work on flash photolysis to study fast chemical reactions, George Porter was both an outstanding Director of the RI and the Davy–Faraday Research Laboratory (DFRL) and an outstanding Fullerian Professor of Chemistry there. He was particularly proud to be successor at the RI of Sir Lawrence Bragg, who, according to Porter, told him on the day he became Director: *'The RI is like a precious egg: handle it very carefully.'*

Writing in the *New Scientist* in September 1977,[1] on the history and future of the RI, Porter made several important points. Here are some of his statements:

Albemarle Street: Portraits, Personalities, and Presentations at the Royal Institution. John Meurig Thomas,
Oxford University Press. © Sir John Meurig Thomas 2021.
DOI: 10.1093/oso/9780192898005.003.0012

Table 12.1 *Some of the Discoveries Made at the RI*

Table I   Research inventory for the Royal Institution

The kinetic theory of heat (Rumford, Davy, Young).

Wave theory and interference of light (Young).

Electrochemistry (Davy). The laws of electrolysis (Faraday).

Safety Lamp, Carbon arc lamp (Davy).

Discovery of elements: Sodium, potassium and impure barium, strontium, calcium, magnesium and boron (Davy). Chlorine (identified as element), iodine (with Ampere) and fluorine (properties induced) (Davy). Argon (Rayleigh with Ramsay).

Electromagnetic induction. The transformer and the dynamo (Faraday).

The Electromagnetic theory of light (Faraday, Maxwell).

Electrostatic induction. Dielectrics. Diamagnetism. Magnetorotation of polarised light. Magnetic permeability. Paramagnetism of oxygen (Faraday).

Liquefaction of gases. Hydrochloric acid (Davy). Chlorine (Davy and Faraday). SO_2, H_2S, N_2O, NH_3, ethylene (Faraday). Oxygen, hydrogen, fluorine. Solid hydrogen (Dewar). Thermos flask (Dewar).

Benzene, isobutylene (Faraday). Naphthalene (Brande, Faraday). Acetylene (E. Davy).

Theory of heat and radiation (Rayleigh).

Light scattering (and the blue of the sky), radiation measurement, glacier flow (Tyndall).

Structure of graphite, organic molecules, minerals, hair, wool and very many other crystalline substances (W. H. Bragg and W. L. Bragg, with Bernal, Astbury and Lonsdale).

Development of the X-ray diffractometer (Arndt and Phillips).

Measurements on haemoglobin and myoglobin (Bragg, with Perutz and Kendrew in Cambridge).

First structure of an enzyme (Phillips).

(Courtesy George Porter)

'*To recite the research work of the institution is to tell the history of a large part of British science and there can be no laboratory of comparable size where such a wealth of discoveries have been made.*' (Table 12.1 shows a few of them.)

'*Today, the Royal Institution serves science in three ways: First, by the advancement of knowledge through research; secondly, by encouraging a greater interest in a better understanding of science; and thirdly, by acting as a trustee of a building of great scientific historical importance. This is a unique combination which does not fit into any government pigeonhole but it is one which has worked admirably and continues to do so.*

'*The logic of this triple alliance of science—history, exposition, research—is compelling. Here, in the middle of London, is the laboratory with the longest history of scientific achievement of any in Britain, and probably of any in the world. It has been not only the laboratory but the home of more great scientists than have lived in any other single house.*'

12.2.1   Important Advice from George Porter

While preparing to move my equipment and some co-workers from Cambridge to London, I had many private discussions with my predecessor. In summary, below is the gist of what he said:

'Bear in mind your triple responsibilities.

'*To advance knowledge (i.e., do the highest quality research).*

'*To encourage a greater interest in and a better understanding of science among members and the general public, as well as among schoolchildren.*

'*To act as a trustee of a building of great scientific-historical importance.'*

He also told me:

'*Never invite anyone to give a Discourse, unless you have heard (and approved of) that person's lecture before.*

'*Insist on emphasizing to the selected Discourse speaker that he or she was expected to write up a version of their Discourse for the Proceedings of the Royal Institution.*[2]

'*Always have ready a Friday Evening Discourse, just in case the speaker fails to turn up or falls ill, and the members assembled in the lecture theatre would be disappointed.*[3]

'*Do as much as you can to enhance the profile and reputation of the RI.*

'*Support all its activities; and always endeavour to boost its financial resources.'*

12.3   Kathleen Lonsdale

As a research student in chemistry at the University College of Wales, Swansea, I was given a problem that required me to know all the key structural features of graphite and diamond. As I had not studied crystallography as an undergraduate, I searched in our well-stocked library for appropriate books on crystals that catered for someone of my primitive knowledge. I remember coming across (Dame) Kathleen Lonsdale's book[4] '*Crystals and X-rays*', published in 1948, which I found very helpful. It enlightened me on the phenomenon of twinning in crystals, and I also came across space groups, which were hitherto unknown to me.

In 1956, my research supervisor took up a Professorship in Queen Mary College, London, and so I and three other research students left Swansea and found accommodation in Central London.[5] On Saturday nights, I would go to play snooker in the University of London Union (in Malet Street). I would play games on the snooker table with anyone who was interested in joining me. One night a student named Steve asked me to play a game of snooker with him. He was a most friendly individual. After several weeks, I asked him his name: '*Lonsdale*', he said. I told him that the only Lonsdale I had encountered previously was the crystallographer: '*That's my mother*', he said.

Fast forward to 1969, when I was Head of the Department of Chemistry at the University College of Wales, Aberystwyth.

As I was doing a good deal of lecturing (often in Welsh) at local village communities on popular science topics, I became enamoured of the idea that a distinguished scientist could be invited to my Department, and to give a popular lecture, not only to university students, but to local people (schoolteachers, doctors, local politicians). So I wrote to Kathleen Lonsdale and asked her if she was prepared to come to Aberystwyth under the circumstances. (In my letter, I asked her to convey my regards to her son, Steve). She immediately responded and her lecture, to a packed theatre, was a great success. She even told the audience how many grandchildren she had. She also talked a great deal about Sir William Henry Bragg and her days at the DFRL (see Figure 12.1), where she was the first scientist to prove that benzene was flat, and she became the first woman to be elected Fellow of the Royal Society. She also went to prison for a while because she was a pacifist; she was the first woman to serve as President of the British Association; she was an authority on diamond and on urinary calculi; and she was an altruistic person, judging by the way she treated me, my wife, and our three-year-old daughter, Lisa.

Figure 12.1 *From left to right, front row, W. T. Astbury, Kathleen Yardley, W. G. Burgers, J. M. Robertson, and R. E. Gibbs; back row, Eric Holmes and Boris Orelkin. This is the research team that William Henry Bragg had assembled at the Davy–Faraday Research Laboratory (DFRL) in the late 1920s.*
(Courtesy Royal Institution (RI))

Kathleen stayed an extra day in Aberystwyth and took a great interest in my group's studies of photochemical reactions in the organic solid state. (She communicated a paper[6] by my colleagues and myself to the Royal Society, and recommended my work to be exhibited (as it was in due course) in a future summer Soirée.) She told me a great deal about the happy atmosphere that pervaded the DFRL in her days. I subsequently corresponded with her monthly, up until the time she died in 1971. She gave me descriptions of how she felt, and also of the medicines she was prescribed.

In later years, I became friendly with several of her ex-graduate or postdoctoral students, notably Professor Mike Glazer and Sir Ronald Mason. Mike Glazer[7,8] told me that she was a superb crystallographer, and the best-ever scientist that he had ever encountered in reading the significance of Laue X-ray patterns (such as is typified in Chapter 11).

12.4   Michael Atiyah: Knots at the RI

When Sir Michael Atiyah passed away in 2019, *The Times* and other newspapers in the UK and US said in their obituaries of him that he was reckoned by most experts in the world to be the greatest mathematician since Isaac Newton. Like Newton, he held the post of President of the Royal Society (1990–95). He was also a Field's Medalist and he won the Abel Prize, arguably the highest honour that can be accorded to a mathematician. He had also been Master of Trinity College, Cambridge, and Head of the Isaac Newton Institute.

I first met him when we both sat on the Main Committee of the UK Science Research Council (1986–91). His cogent interventions and suggestions to the Chairman were always given at top speed, and they usually commended themselves to other members of the Committee immediately.

But, why do I include him in my RI reminiscences? It all started when I approached another spectacular Cambridge mathematician, Sir Peter Swinnerton-Dyer, in 1987, asking him if he would be prepared to present a Discourse at the RI on a mathematical topic that would interest members and their guests. He immediately responded that he would be inferior in doing this compared with Michael Atiyah, whom he said would not only illuminate pretty well any mathematical topic of his choice, but that also he would entertain the audience, as he was such a spectacularly successful lecturer. Atiyah duly came to give his Discourse in 1988 on 'The Geometry and Physics of Knots'. This is how he began:

'When mathematicians address a general audience, even as enlightened as attends a Royal Institution Discourse, he faces a daunting task. Mathematics can be such a highly technical and abstract subject that communicating its latest developments to a lay public presents formidable difficulties. Bearing this in mind, I selected this evening's topic according to the following criteria. It should have: (1) a simple visual content; (2) an interesting historical background; (3) a relation to physics; (4) a recent exciting discovery.'

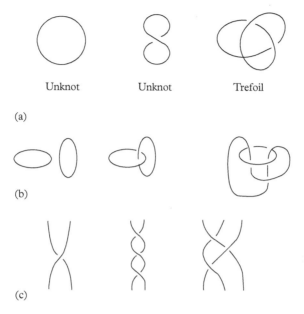

Unknot   Unknot   Trefoil

(a)

(b)

(c)

Figure 12.2 *Illustrations of (a) knots, (b) links, and (c) braids from Sir Michael Atiyah's Discourse on the geometry and physics of knots.*
(Courtesy the RI)

Sir Michael then proceeded to fascinate his audience, first by describing the nature of knots, links, and braids (see Figure 12.2).

A link, he said, is like a knot except that it is made up of several closed pieces of string. Finally, a braid is a collection of strings or 'strands' which may be entangled, but which all move in the same direction. Examples of braids are shown in Figure 12.2 (c) above.

Knots had attracted many of our antecedents, as in the classical story of Alexander the Great and the Gordian Knot. But, in Lord Kelvin's days (at the RI and elsewhere), there was talk of 'vortex atoms', an idea put forward by Rebein in 1867. At that time, the ultimate nature of matter was a mystery (*'it still is'*, said Atiyah!), and Kelvin had the magnificent idea that atoms consist of *'knotted vortex tubes of the ether'*, because of their stability, vibrations, and transmutation. Kelvin and P. G. Tait, one of his collaborators, spent ten years studying and calculating knots (see Figure 12.3).[9]

Atiyah went on to describe the so-called invariance of knots and then went on to the topic of topology (which is, in essence, a subject on which Tait became the prime authority; it is concerned with the science of connectedness).

Despite the great advances made in topology[10] in the latter part of the twentieth century, progress in knot theory has been slow. It came as a big surprise to expert mathematicians such as Atiyah in Cambridge and Edward Witten[11] in

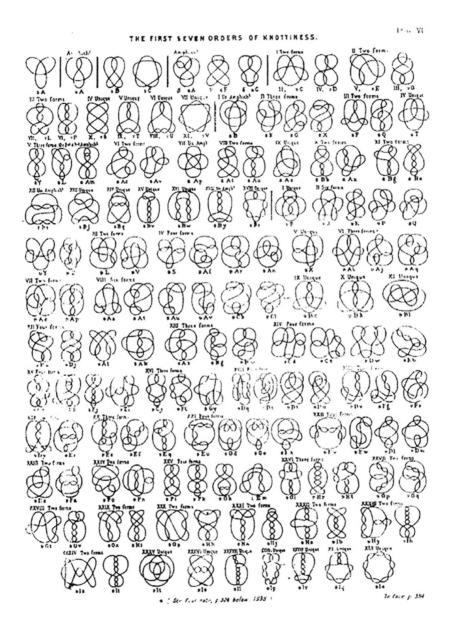

Figure 12.3 *A diagram from one of Lord Kelvin's publications, quoted by Michael Atiyah, 'The First Seven Orders of Knottiness'.*

(Courtesy the RI)

Princeton when the New Zealand mathematician Vaughan Jones discovered what is called the Jones polynomial, dealing with the concept of the invariants of knots.

The Jones polynomial is too advanced a subject to describe here, but it has many remarkable properties; and these have been described in Chapter 1, Section 1.7. The Jones polynomial, *inter alia*, has led to the proof of many of Tait's century-old conjectures. More important, it has huge repercussions in physics, quantum mechanics, and even in biological structures composed of DNA.

It is interesting that knots are now appearing increasingly in biology—see the work of Professor Sophie Jackson (Figure 12.4).

12.4.1   The Responsibilities of Scientists at the Royal Society

In November 1992, Michael Atiyah gave his Anniversary address as President of the Royal Society. His remarks were greatly appreciated by most as the voice of a sane, concerned scientist. However, a few hawkish Fellows of the Royal Society did not care very much for what he said. I consider many of his remarks to be highly relevant, and consequently some of them are repeated here:

'*Science, and the technology that flows from it, now permeate all aspects of modern life. To such an audience as this it is hardly necessary to dwell on the extent to which science has transformed our economy, our health, our environment and our military capabilities.*

Figure 12.4 *Professor Sophie Jackson is holding a knot—this is the knot that is formed by the enzyme ubiquitin C terminal hydrolase.*[12]
(Courtesy Sophie Jackson)

The process has been going on for several centuries, but it has accelerated enormously in the 20th century and there are profound questions over where it is going to take us in the 21st century. A few generations back all these changes were described as "progress", and there was a certain naive optimism implied by that word. Things were getting better thanks to science and there was a rosy future ahead. With this complacent scenario science had a clear conscience, and the Royal Society an easy task.

'*Until the Second World War, opposition to scientific progress was a minority view exemplified by the romantic movement in the arts. Nostalgia for a mythical pastoral paradise was hardly something to be taken seriously.*

'*All this changed, literally with a bang, in 1945. The atomic bomb ushered in an era in which negative aspects of science have become increasingly prominent. The scientific conscience has had to grapple with serious questions, and a strong anti-science lobby has grown up which attacks the fundamental premises of scientific research. Interestingly enough, the very word "progress" has almost disappeared from our vocabulary and has been replaced by the more innocuous word "growth".*

'*The conventional answer of scientists to these developments has been to argue that science per se is morally neutral and that it is only the applications of science, as determined by those with political or economic power, that can have socially negative consequences. While logically impeccable this defence is not really adequate. One might as well argue that the arms dealer is not really responsible for the lethal use to which his weapons are ultimately put.*

'*Scientists with a more developed conscience do, I think, recognize that, as the ultimate masters of the genie in the bottle, they have a responsibility to inform, warn and advise on the uses of science. Moreover to be effective this may involve scientists, and scientific organizations, in participating actively in public debate and in the political process.*

Figure 12.5 *Sir Michael Atiyah OM.*
(Courtesy Trinity College, Cambridge)

'*If we look back at the 20th century, which is now drawing to a close, we can see that it was dominated by the two World Wars and the extended Cold War that succeeded them. Besides the tragedy of human suffering, vast resources, much of it science-based, were expended on armaments. In particular, large parts of the world diverted resources which they could ill afford for the purchase of expensive high-technology weaponry provided by the more scientifically-advanced countries.*

'*Fortunately a nuclear holocaust has so far been avoided and with the end of the Cold War this threat is receding rapidly, though the safe dismantling of the nuclear arsenals will take time and require careful monitoring. The scientific community has a part to play in all this and I would like at this stage to pay tribute to those scientists, including Fellows of the Society, who through the Pugwash movement and other bodies have helped over many difficult years to introduce a measure of sanity into an otherwise insane process.*

'*Just as the nuclear shadow is passing, it is in danger of being replaced by the more diffuse but no less devastating threat of chemical and biological warfare. This threat is much more difficult to monitor and control.*

'*Large expensive installations are not necessary and it is hard to distinguish between laboratories designed for civilian or military research. The basic scientific information will inevitably spread, and it is only a matter of time before the potential for chemical and biological warfare is widely available. Strict security on sensitive information may provide a breathing space, but it cannot be relied on in the long run. The only real security lies in resolving the underlying causes of conflict and in establishing the appropriate political structures. We have perhaps a few decades in which to tackle these fundamental problems.*

'*While weapons of mass annihilation are the major threat, conventional modern weapons are nasty enough and it is depressing to see that, even as the Cold War ends and signs of sanity appear in some of the world's trouble-spots, the arms trade seems to be as active as ever. As long as scarce resources are lavished on extravagant military hardware the real economic and political problems of the world will continue to fester.*'

12.5   Margaret Gowing

Margaret Gowing, the only female historian of science who has been elected to both the British Academy and to the Royal Society (emulating a few male academics, such as Joseph Needham and Karl Popper), presented an exciting RI Discourse, when she addressed the subject of Niels Bohr, the great Danish physicist and pioneer of nuclear physics in the 1940s. My wife and I became good friends of hers, after we spent two fascinating hours talking to her in the Director's flat, at the completion of her Discourse, and when all the other guests had departed (see Section 12.14).

According to one of the many obituaries written on her, Margaret Gowing, who came from a working-class background, won a scholarship to the London School of Economics, where she was much influenced by the versatile and impressive economic historian Eileen Power, who was at once a distinguished historian and a redoubtable champion of a variety of causes that reflected her keen perception of what constituted the public interest.

Margaret Gowing's scholarly reputation rested primarily on her magisterial studies of atomic energy in the UK during and after the Second World War. My debts to her are twofold. First, at lunches taken in the Royal Society, when I was domiciled in Albemarle Street, she introduced me to three eminent individuals: Sir Rudolf Peierls (co-author of the Frisch—Peierls Memorandum of 1940[13] that ultimately led to the formation of the Manhattan Project in World War II); Professor Nicholas Kurti, the British-Hungarian low-temperature physicist and activist; and Lord Sherfield, former Ambassador to the US and also former Chairman of the Governing Body of Imperial College and Chairman of the UK Atomic Energy Authority from 1960—64. (Lord Sherfield gave a Discourse at the RI in 1988 on *'The Science and Technology of Great Britain'*).

Margaret Gowing fell in love with the RI while I was there, and became deeply involved in knowing all about the career of Michael Faraday. She was full of excitement when she entered 21 Albemarle Street; the sight of a coal fire within a few yards of the front door prompted her to comment that this was the sight when early members of the RI came to listen to the lecture-demonstrations of Davy and Faraday. Partly because of her almost juvenile enthusiasm, she expressed the desire to help me as Director. This is my second debt. In fact, she volunteered to read every word of the draft of my short monograph on Faraday when I wrote it in the early summer of 1991.

Figure 12.6 *Margaret Gowing, the only historian of science to be elected Fellow of both the British Academy and the Royal Society.*

(Wikipedia open access)

What was gratifying about her approach, as a fresh and new breed of historian of science, was that she focused more on Faraday's achievements rather than on speculative philosophical aspects that, according to some, greatly influence the science carried out in a particular age. This attitude of hers was much admired by the electors to her Oxford University Chair: Lord Dacre, Sir Rudolf Peierls, Nicholas Kurti, and Lord Dainton. This characteristic made her a refreshing critic of whatever I used to say in my various articles about science in general, and Faraday, Davy, and Rumford in particular.

12.6   Some of my Activities and Duties During the Faraday Bicentenary

During the bicentenary celebrations of the birth of Michael Faraday, as Director, I was expected to do very many things and to undertake several unusual tasks.

I was successful in persuading the Bank of England to place Faraday (at his RI bench in characteristic stance) on the £20 note (see Figure 1.10, in Chapter 1). It was also possible to persuade the General Post Office to issue a special first-day stamp. Further, I also persuaded the Science Museum to put on an exhibition of some of Faraday's work; and the Royal Society Soirée that summer contained a number of Faraday's exhibits and experiments. Sir Brian Pippard (our Visiting Professor in Physics) and I also mounted a special exhibition on *'Faraday and his Contemporaries'* at the National Portrait Gallery (see Figure 12.7). We wrote the handlist, and I gave a lunchtime lecture-demonstration on the *'Genius of Michael Faraday'*. The special Westminster Abbey Memorial Service (in September 1991) has already been described in Chapter 4. There were numerous other occasions, involving the Royal Society of Chemistry, the sesquicentenary of whose formation was also in 1991, where I gave lectures on Faraday. I spoke on this topic, in all, thirty-one times in the UK, US, and India.

My lecture in the National Physical Laboratory, New Delhi, on Faraday was the K. S. Krishnan[14] Memorial Lecture. While in India, I was scheduled to give several Faraday-based lectures in Bangalore, Mumbai, Chennai, Pune, Khanpur, and elsewhere. However, shortly before my wife and I were due to leave Delhi for the other venues, the Gulf War broke out. I immediately contacted the British Embassy, who advised us to return as soon as possible to the UK, as India received her oil from Kuwait, and there were fears that no planes would be able to leave India within a few days time. So we returned to London.

12.6.1 An opportunity to write a Book on Faraday

I had numerous facts about Faraday at my fingertips in view of what I had intended to do in India. The RI had given me leave to be away for some three weeks or so. My duties as Director at the RI were to be undertaken by my able Deputy, Professor Richard Catlow (later of University College, London, and Cardiff University,

MICHAEL FARADAY AND HIS CONTEMPORARIES

Professor Sir John Meurig Thomas FRS
Professor Sir Brian Pippard FRS

A BRIEF BIOGRAPHY

Michael Faraday, arguably the greatest experimental scientist ever, was born on 22 September 1791 in Newington Butts, Surrey, now Elephant and Castle in London. His ailing father, a blacksmith from Yorkshire, died when Faraday was in his teens. After an elementary education, consisting in his own words of 'little more than the rudiments of reading, writing and arithmetic', he left school at the age of thirteen and worked first as a newspaper boy and then as an apprentice learning the arts of bookbinder, stationer and bookseller from a Mr Riebau in Blandford Street. He took a keen interest in the contents of the books, especially the scientific ones, and 'made such simple experiments in chemistry as could be defrayed in their expense by a few pence per week, and also constructed an electrical machine'.

Figure 12.7 *Opening paragraph of the handlist for the exhibition at the National Portrait Gallery, 1991, organized by Sir John Meurig Thomas and Sir Brian Pippard. (Copyright John Meurig Thomas and Brian Pippard)*

and also Foreign Secretary to the Royal Society), so I took advantage of this 'unexpected leave' to write a popular book on Faraday. Working more than twelve hours per day, I was able to access all (or essentially all) the original papers, letters, and manuscripts of Faraday's with the great help of the Librarian and Archivist, Irene McCabe, and my wife, Margaret, I was able to meet the deadline set for me by the Institute of Physics. The book appeared just before I ceased to be Director of the RI. It contained an elegant and very rapidly written Foreword by Brian Pippard. I was happy to receive from the White House a

Figure 12.8 *'Michael Faraday and the Royal Institution: The Genius of Man and Place'.* *(Courtesy the Institute of Physics)*[15]

letter from the US President's Scientific Adviser acknowledging the importance of the story it told, and later to see the book translated into Japanese, Italian, and Chinese.

12.7 Brian Pippard

Shortly after my appointment to the Directorship of the RI, I phoned up Brian Pippard, who had just taken early retirement as Cavendish Professor at the Department of Physics, Cambridge, but who still lived an extremely active life as a lecturer, brilliant musician, and versatile educator (to science societies in the colleges of Cambridge). I invited him to become an RI Visiting Professor of Physics, in the tradition of his predecessors as Cavendish Professors (James Clerk Maxwell, Lord Rayleigh, J. J. Thomson, Ernest Rutherford, and Sir Lawrence Bragg). He immediately accepted, and expressed great willingness to give Discourses, lunchtime lectures, and assist greatly in the activities associated with the Faraday bicentenary.

Brian Pippard[16] was one of the most brilliantly able persons that I have ever met. Not only was he an outstanding experimental physicist (whose graduate student Brian Josephson won the Nobel Prize at the early age of thirty-three), he was also extremely articulate and witty in the spoken word. And his skills and style as a writer were outstanding. The speed with which he could write up a complicated account of any scientific (numerical or social topic) was remarkable. This is well-illustrated by the joint work we carried out at the Portrait Gallery, when we mounted the bicentenary '*Faraday and his Contemporaries*' exhibition on Michael Faraday (see Figure 12.7).

We each agreed that for the narrative account required for the exhibition hand-list, I would cover the chemical contributions and discoveries of Faraday, and he the physics equivalent. I took three whole days, involving several revisions of my text. He did it in one night's sitting, and the manuscript he handed me required no revision.

I was particularly happy to have Brian as a Visiting Professor at the RI for another reason. I had often bemoaned (to him) the fact that so many of our distinguished speakers at the RI seldom made full use of the huge dimensions of our unique lecture theatre, especially its enormous headroom. (There was an exception when David Phillips revealed the amino acid sequence (see Chapter 1, Figure 1.7) of the 129 residues, appearing from the ceiling in his memorable lecture on the structure of lysozyme.)

Brian Pippard agreed to a suggestion of mine that he should re-enact the dramatic experiment that the French physicist Léon Foucault had carried out in the Panthéon, in Paris, in 1851, when he demonstrated so vividly the rotation of the Earth, by the change of plane of the pendulum's oscillation (see Figure 12.9).

 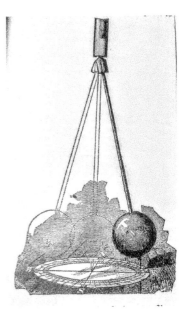

Figure 12.9 *Illustration of the historic demonstration by Léon Foucault (at the Pantheon, Paris) of the rotation of the Earth from the movements of his famous pendulum.*
(Wikipedia open access)

Figure 12.10 *Brian Pippard with the so-called parametrically driven Foucault pendulum that he designed and had built for the Science Museum.*
(Courtesy of the Science and Society Picture Library, London Science Museum)

In his famous Discourse at the RI, Brian Pippard set up his point of suspension of his pendulum, some fifty feet above ground level. In the duration of the lecture (one hour), the reality of the rotation of the Earth could be readily perceived (see Figure 12.10).

12.8   My First Discourse Speakers

Conscious of the advice given me by George Porter concerning how I should select Discourse speakers (Section 12.2.1.), I made sure that I had absolute certainty about the high quality of the work and communication skills of the chosen speakers (see Table 12.2).

All these Discourses drew large audiences, and the non-scientists in the audiences expressed satisfaction concerning the level of information that the speakers assumed in conveying their disparate messages.

My own inaugural Discourse dealt with a subject close to my heart: the poetry of science. For several years, from my time in Aberystwyth, I had spoken on this topic in English and Welsh, to many societies in Wales, and to the U3A audience soon after it was established at Cambridge.

Table 12.2 *Invited Discourse Speakers*

Speaker	Date	Institution	Discourse Title
Professor Jacques Heyman	17 October 1986	Head of Department of Engineering, University of Cambridge	'*Structural Analysis of Gothic Architecture*'
Professor Raymond Andrew	24 October 1986	Department of Physics, Florida State University	'*Magnetic Resonance Imaging: Seeing Safely Inside the Human Body*'
Professor Colin J. Humphreys	31 October 1986	Department of Materials Science and Metallurgy, University of Cambridge	'*Dating New Testament Events Using Astronomy*'
Professor J. M. Thomas	7 November 1986	Director of the RI	'*The Poetry of Science*'
Professor Paul C. W. Dennis	14 November 1986	Department of Physics, University of Newcastle	'*What Wound up the Universe?*'
Professor Roald Hoffmann	21 November 1986	Nobel Laureate, Baker Laboratory, Cornel University	'*The Logical Structure of Modern Chemistry —or What Chemists Really Do*'
Professor Margaret Gowing	28 November 1986	Department of History of Science, University of Oxford	'*Niels Bohr and Nuclear Weapons*'
Professor Sir John Butterfield	5 December 1986	Reguis Professor of Physics, University of Cambridge	'*How Unhealthy are the British?*'

My Discourse on 'The Poetry of Science' began:

'During the course of his Reith Lectures on "Science and the Common Understanding" broadcast by the BBC in the autumn of 1953, J. Robert Oppenheimer remarked, tongue in cheek perhaps, that:

"Physics seeks to explain in a simple language something that nobody knows: poetry seeks to say things that everybody knows in a language which nobody understands."

'To an impressionable undergraduate, which I was in those distant days, Oppenheimer's words possessed a meretricious quality, which soon died away upon reflection. His description of physics was a tolerable approximation, but I questioned his views on poetry. Indeed, in that youthful autumn, I grew convinced that he was wrong when I read in the London Times an extended obituary which recalled passages of a poem describing a farmstead that lies a few miles away from the village in South Wales where I was born:

"And as I was green and carefree, famous among the barns
About the happy yard and singing as the farm was home,
In the sun that is young once only
Time let me play and be,
Golden in the mercy of his means"

Another poem of Dylan Thomas, quoted in that Times obituary, was written when the poet was himself a teenager:

"The force that through the green fuse drives the flower
Drives my green age; that blasts the roots of trees is my destroyer
And I am dumb to tell the crooked rose
My youth is bent by the same wintery fever"

The magic and message of that verse, along with the imagery and the rhythm of the words, are irresistible. Without wishing pedantically to pursue definition, one recognises this as true poetry. Just like physics, and science generally, poetry is, amongst other things, a valid means of communication; and is the product of a creative act. Furthermore, it is not, in my opinion, correct to argue, as did Robert Frost, that "Poetry is what gets left behind in translation," for we know from the literature and culture of other nations, that the greatest poetry, like the greatest science, transcends the original medium in which it appears.

'Rabindranath Tagore, wrote most of his poems in his native Bengali. You may recall that his "Gitangali" was composed not long after the death of his wife, son and daughter.

"Thou has made me endless, such is thy pleasure.
This frail vessel thou emptiest again and again
And fillest ever with fresh life.
This little flute of a reed thou has carried over hills and dales,
And has breathed through it melodies eternally new"

Now I propose to spend little time this evening citing poetry as such. Instead I shall illustrate visually, and necessarily somewhat superficially, why I espouse Wordsworth's dictum that:

"Poetry is the impassioned expression which is the countenance of all science" and why T. S. Eliot's thesis rings true that poetry concerns itself with…"The element of surprise and elevation of a new experience".

'Now science, in any of its manifestations, cannot, it seems to me, reach the depths of human psyche and emotion in quite the way that poetry and music can. A medical or scientific statement of death, for example, hardly rivals the impact of the poetic declaration:

"When once our short life has burnt away, death is an unending sleep."

Nevertheless, science has an indisputable aesthetic dimension, and it can fill us with awe and inspiration; and through its pursuit we can be suffused with an appreciation of beauty, elegance and mystery.

"The most beautiful experience we can have is the mysterious. It is the fundamental emotion which stands at the cradle of true art and true science. Whoever does not know it and can no longer wonder, no longer marvel, is as good as dead, and his eyes are dimmed" (Albert. Einstein, 1934). Einstein also said: "The most imcomprehensible fact of nature is the fact that nature is so comprehensible," (The late Frederick Seitz, President Emeritus of Rockefeller University, once drew my attention to the following remark by James Clerk Maxwell (1860): *'We shall find that it is the peculiar function of physical science to lead us...to the confines of the incomprehensible.')*

'It is now widely accepted that the scientist's discoveries first emerge from imaginative leaps and intuitive insights: he then proceeds to prove by a process that entails logical analysis and repeatable experiment.

'But, the joy, romance and poetry of science comes from its intrinsic excitement, sweep and incompleteness and from the surprises and pleasures inextricably mingled with the correlation of seemingly disparate phenomena.'

On creativity and imagination in science, I continued:

'The scientist composes poems not with words but with models and concepts, with adumbrations, analogies, syntheses and sketches, with photographs and equations, and with new techniques and new designs, all products of his insight. In common with the poet he is frequently driven by an obsessional zeal and a touch of neurotic punctilio. In processing from a confusion of understanding to a new understanding of his confusion, his mood can move from saturnine gloom to sublime elation.

'Certain scientists, in addition to their skills as scientists, also fashion their expression in poetic ways. Davy and Faraday, as you will have been reminded by our exhibition in the Library this evening, were masters in this regard. I will cite a favourite example from a widely read author in my own subject. Picture him ruminating over the distribution in kinetic energies possessed by molecules in the gaseous state, questions that had been discussed quantitatively and elegantly by his predecessors Maxwell and Boltzmann in the last century.

"Energy among molecules is like money among men. The rich are few, the poor numerous."

(It would not surprise you, I am sure, judging by those moral overtones—*'like money among men'*—that the author of those words, the late Dr Moelwyn-Hughes, of Cambridge, was the son of a Welsh hymn writer!)

'I spoke a moment ago about discovery emanating from imaginative leaps. This happens in all fields of science, and doubtless in many other enterprises. In mathematics, theorems are often advanced without proof and, in the fullness of time, are shown to be true: but the mathematician who first formulated the theorem may not, himself or herself, have provided a defensible logical demonstration of its veracity. This ability to divine the truth stimulates the creative skills of musical geniuses like Mozart.

$$1 - 5\left(\frac{1}{2}\right)^3 + 9\left(\frac{1.3}{2.4}\right)^3 - 13\left(\frac{1.3.5}{2.4.6}\right)^3 + \ldots$$

$$= \frac{2}{\pi}$$

1729

$$635318657 = 134^4 + 133^4$$
$$= 158^4 + 59^4$$

Figure 12.11 *Ramanujan and some of his numerical insights. The first number that can be constructed from the sum of the cubes of two other numbers in two different ways is 1729. On being told by G. H. Hardy, who visited him at his hospital bedside, that this was the number of the taxi in which Hardy had travelled, Ramanujan immediately stated the above result. When Hardy asked him what was the smallest number that could be written down as the sum of two fourth powers in two ways, he replied* 'a very large number'. *It is the number 635318657, worked out by my colleague Professor J. Klinowski.*
(Copyright John Meurig Thomas)

'One of the most fascinating characters in twentieth century mathematics was Ramanujan, who, when he wrote in 1913 his famous letter to G. H. Hardy, the renowned mathematician at Trinity College, Cambridge, was employed as a clerk in the docks of Madras.

'Several theorems appeared in Ramanujan's letter. Some, in the words of Hardy, appeared wild and fantastic; one or two were already well known. To cut a long story short, Hardy resolved to bring Ramanujan to Cambridge. He succeeded. Ramanujan became a Fellow of Trinity and shortly thereafter a Fellow of the Royal Society.

'Hardy's comment on some of Ramanujan's theorems was: "They must be true because, if they were not true, no one would have had the imagination to invent them." Here, we indeed enter a strange realm, where, by implication, even the imagination, at least for ordinary mortals, is not enough to arrive at the truth.'

12.9 The Night of the Monarch Butterfly

As a result of reading a remarkable article in *The National Geographic Society Magazine* in 1976, I became fascinated by the monarch butterfly. Among other things, these beautiful orange-coloured creatures, it was stated, gathered in their tens of millions, filling the sky with movement for as far as the eye could see above a primeval fir forest thousands of feet up in the mountains of Central Mexico. This unusual sight has lasted tens of thousands of years, and, so I recalled, something like three-hundred million of the monarch butterflies 'disappear' each year from Canada and the US as they migrate to Mexico.

One evening, prior to a Discourse, Sir William Harding, the former UK Ambassador to Brazil, told me that the husband of his niece was an expert on the monarch butterfly. He further declared that this man gave a 'wonderful' lecture describing the life cycle and other aspects of monarch butterflies. While blatantly ignoring the advice given to me by George Porter (see Section 12.2.2 above), I immediately told Sir William that I would invite his relative to come and present a Discourse at the RI. (Only on one other occasion did I ignore the advice of George Porter regarding choice of RI speakers—see Section 12.11 below.)

My habit on Fridays was to arrange to meet that evening's Discourse speaker just outside the Bernard Sunley Lecture Theatre close to the entrance to the RI at *ca.* about 12.20 pm, after the end of my research group's weekly seminar in the Bernard Sunley.

After exchanging greetings about safe journeys, comfortable hotels, etc., I would ask the speaker how many slides would be shown during the Discourse, as this gave me an accurate sense of how things would turn out that evening. The speaker, on 3 November 1989, Dr Carlos F. Gottfried (President of Monarco AC of Mexico), replied, *'I shall be showing eighty in the first part of my talk and 102 in the second.'* On hearing this, I very nearly had apoplexy! *'You know that you have only an hour?'* He responded, *'Yes, I do, and I timed my talk with my wife's help in our hotel last night, and I shall finish with about one minute to spare.'* I asked, with obvious concern, *'Have you given this talk before?'.* To this he replied, *'Only in Spanish'*, which intensified my concern. *'By the way'*, the speaker continued, *'I shall be playing J. S. Bach's Wachet Auf in the final minutes of my presentation.'*

Up until the time of the pre-Discourse dinner in the Director's flat, I grew increasingly concerned that his Discourse could either be a catastrophic failure or a triumphant success. Mercifully, it was the latter; and when Bach's music harmonized beautifully with the rapid fire of slides of the migrating butterflies, goose pimples covered by arms, and my hair seemed to stand on end. It was a brilliant

Figure 12.12 *Postage stamps illustrated with the monarch butterfly.*
(Copyright John Meurig Thomas)

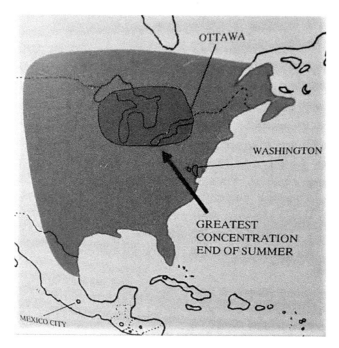

Figure 12.13 *The locations of the monarch butterflies at the end of the summer, which are around the Great Lakes in the US and Canada.*
(Public domain)

performance, and I thought his lecture was riveting. (It is elegantly described in the *Proceedings of the RI for 1990*.)[18]

Gottfried masterfully reviewed several aspects (and hitherto unsolved mysteries) of the monarch's life history and its entire continental migration characteristics, which involve up to five generations to complete.

12.9.1   A Human Being Weighing Nine Tons!

One devastatingly interesting fact that emerged from the published version of Gottfried's Discourse, which was not mentioned in his live version, concerns the way in which the monarch butterfly emerges from its minute beginnings. The larvae go through five different stages of growth, or so-called instars, denoted at the end of each one by the moulting of their skin. According to Gottfried:[18]

'Moulting in insects is carefully and exactly controlled by hormones...The complete development and growth period takes approximately 15 days. The larva, when it eats its way out of the eggshell weighs approximately 0.5 mg. At the end of 15 days it weighs 1,500 mg, having multiplied its weight 2,700 times. (A normal human body with a weight of approximately 8 lbs growing at the rate of the caterpillar would weigh over

22,000 lbs, i.e. 1,600 stone, or nine tons in 15 days.) No other class of animals matches this fantastic growth rate.'

12.10　Neil MacGregor

It was during a conversation that I had in 1989 with my fellow physical chemist Sir Rex Richards, former Head of the Physical Chemistry Laboratory and former Vice-Chancellor of the University of Oxford, that it was suggested to me that I should invite the then Director of the National Gallery, Dr Neil MacGregor, to give an RI Discourse. Not having heard him speak, I interrogated Rex Richards further. He told me that, in his capacity as a member of the Scientific Advisory Committee of the National Gallery, he had heard its Director give brilliant talks to its Trustees, in a peripatetic fashion, while moving from one portrait or scene in the Gallery to another. So this is how I came, for the second time, to disobey George Porter's advice.

Neil MacGregor, now one of the world's foremost art historians, studied modern languages and philosophy (in Oxford and Paris respectively) before being persuaded by Anthony Blunt to study art history under his supervision at the Courtauld Institute, University of London. Blunt said of MacGregor that he was the most brilliant student he ever taught.

On 4 May 1990, Neil MacGregor gave his excellent Discourse on *'Science and the National Gallery'*. His abstract was as follows:

'To conserve a painting you must know what it is made of. To understand how a painting was originally intended to appear, you must know what changes are likely to have occurred later in its pigments. To prevent further changes you must be able to

Figure 12.14　*Neil MacGregor, OM, Director of the National Gallery, London, from 1987–2002. (Wikipedia open access)*

control its environment. In short, for the National Gallery to function properly, it is essential that it has the best possible scientific advice. In the last century, it turned to Faraday—among others—and for the last fifty years it has had its own Scientific Department.

'*This Discourse will look at the ways in which the work of the Gallery's Scientific Department has changed art historians' approach to European painting from the Renaissance to the twentieth century. The ability confidently to distinguish between original and later paint has been central to the cleaning of the Gallery's pictures and the recovery of their original impact. With infrared reflectography it has been possible to reconstruct the preliminary drawing beneath the paint surface, thus to come closer to the artist in the very process of making up his mind. And as we discover the making of the picture, the achievements of the artists seem more than ever remarkable*'.

MacGregor's Discourse was outstanding. As well as covering thoroughly the philosophy given in his abstract, he enlightened members of the RI about the subtle differences between Monet and Manet and several other artists with whom they were familiar. (I also distinctly recall his showing a large lorry carrying precious pictures from the Gallery to the empty slate quarries of North Wales to be protected during World War II.[19])

MacGregor's stance towards charging for entry into galleries won great sympathy when Prime Minister Margaret Thatcher required all museums to do so. He resolutely refused to levy charges for entry to the National Gallery. And when he was later invited to become the head of a major New York City gallery, he declined the offer because the practice of that gallery was to charge for entry.

12.11   Three Nobel Laureates

Of the hundred or so speakers that I invited to present Discourses at the RI, one, the phenomenal Roald Hoffmann, with whom I had interacted, first at Aberystwyth and then in Cornell University, was already a winner of the Nobel Prize.[20] He is also a poet, dramatist, and lucid educator. The other two were: the Egyptian-born successor, at the California Institute of Technology, of Linus Pauling (who had given two Discourses at the RI in his early years), Ahmed Zewail; and the lucidly intelligent French physicist Pierre-Gilles de Gennes, of the College de France and Director of École De Physique et Chimie. De Gennes was also an Honorary Member of the RI.

De Gennes, who won his Nobel Prize in Physics in 1991, was a mesmeric lecturer. His Discourse on '*Wetting and Drying*' in March 1990 was a splendid affair. His abstract read:

'*How does water creep in dry soil? At what speed can we spread an ink, a paint, a glue? How fast does a wet plate dry? Many practical questions of that type—and industrial problems—depend on very delicate processes taking place near the "contact time", where the liquid, the solid and air, meet. These processes have resisted our understanding for a long time. Recently, three key experiments have been performed, and a simple global picture has emerged: it will be described here.*'

In his actual Discourse, he began by reminding his audience that one of his forebears was English and, for that and other reasons, he felt privileged to be at the RI. When he came to a critical stage in his Discourse, at the point where seemingly unresolvable problems were encountered, he turned to the audience and said, *'Now, if I were an Englishman, and was confronted with this problem, I would probably say: "I am non-plussed!"'*…This caused the audience to erupt with amusement.

Ahmed Zewail, who won the Nobel Prize in Chemistry outright in 1999, gave his RI Discourse on *'Filming in a Millionth of a Billionth of a Second'* on 27 March 1991. He started by referring to the pioneering, high-speed photography of Eadweard Muybridge (who had given an RI Discourse almost a century earlier), and his elucidation of the dynamics of a horse's gallop. Zewail's work has been described earlier in this book (see Section 11.2.1). When he entered the packed-to-capacity RI lecture theatre, he was taken a little aback, and said, while surveying the audience, *'You must have been told that the speaker this evening was to be another Egyptian: Omar Sharif!.'*

He proceeded to uncover how, with ultra-fast lasers, he and his group in Pasadena, California, were able to record the bond distances and atomic movements on species that 'lived' for no more than 10 fs (ten femtoseconds, 10^{-14}s). This had never been achieved by anyone else previously. Not only was Zewail rewarded with the Nobel Prize, he was festooned with other distinctions up until his untimely death in 2016.[21]

12.12   Officers, Professors, Calendars

Figure 12.15 shows a typical set of events that took place in the May—June period of 1990, my last full year as Director. It can be seen that, following Neil MacGregor's stellar performance, an unusually interesting Discourse was given a week later by one of the creators of the BBC broadcasts *Yes, Minister* and *Yes, Prime Minister*, Sir Antony Jay, who told members about the essence of *'Understanding Laughter'*.

The variety of presentation given in the month of May alone is impressive, with one of the world's experts, the Emeritus Professor and former Director George Porter holding forth at a Public Lunch Hour Lecture on 30 May on the topic *'Solar Energy, Future Prospects'*. Several primary-school lectures were given by the lecturer's assistant and demonstrator Dr Bryson Gore, sponsored by the Shell International Company. And one of the most popular members of the RI in my day, Professor Nicholas Kurti, gave a thoroughly enjoyable and instructive account, with numerous demonstrations of *'Friends and Foes of the Pleasures of Eating'*.

Nicholas Kurti, born in Budapest, was an eminent, low-temperature physicist, who quite early on in his field and in collaboration with his mentor, Sir Francis Simon, both

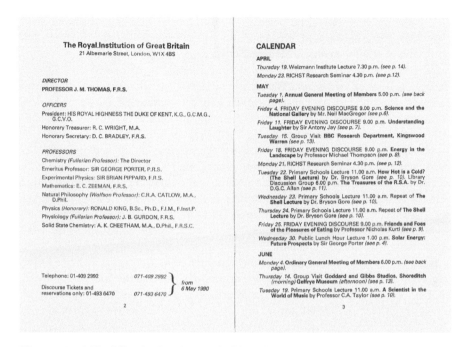

Figure 12.15 *The RI's calendar of events for May—June 1990.*
(Public domain)

of whom had fled Nazi Berlin, came to settle in the Clarendon Laboratory, Oxford. His hobby was cooking, and he was an enthusiastic advocate of applying scientific knowledge to culinary problems (he liked to call himself a 'gastrophysicist'). At a 1969 RI Discourse entitled *'The Physicist in the Kitchen'*, he amazed the members and their guests by using the recently invented microwave oven to make a 'reverse-baked Alaska', otherwise known as 'frozen Florida', (cold outside, hot inside).

The left-hand side of Figure 12.15 shows the names of the Officers and the Professors. I can testify to the importance and efficacy of the President, Treasurer, and Secretary, all of whom did their work voluntarily.

So far as the Professors were concerned, it was invaluable to have George Porter on site periodically. He and I each gave a thirty minute lecture-demonstration to celebrate Faraday's bicentenary. To follow a Porter lecture-demonstration was a daunting task for me. I quoted Shakespeare in my opening words:

'As in a theatre, when a well-graced actor leaves the stage, the eyes of men are idly bent on him that enters next, thinking his prattle to be tedious.'

Professor John Gurdon's outstanding work as Fullerian Professor of Physiology was later to be rewarded by a Nobel Prize in Medicine or Physiology in 2012.

12.13   Deputy Directors and Professorships of Natural Philosophy: 1986–2007

The Professorship of Natural Philosophy at the RI is both of ancient lineage and of great distinction, since it was 'phenomenon' Thomas Young (described in earlier chapters) that was first appointed to this post in 1801 by the founder, Count Rumford.

The Deputy Director's post, as George Porter took great pains to explain to me prior to my moving to Albemarle Street, is also vitally important. The person that holds that post must be one of the best in the country at honouring the three responsibilities expected of the Director: (a) distinction in advancing knowledge; (b) expertise in promulgating scientific knowledge, especially among young people; and (c) a committed custodian of the exceptional historicity of the house of the RI.

It was my great good fortune that the person holding these posts when I arrived in 1986, Professor David Phillips, carried out his duties with conspicuous success.

12.13.1   David Phillips

Professor David Phillips was selected by George Porter as his Deputy Director from among a very strong field of candidates in 1980. He was on the staff of the University of Southampton prior to his translation to Albemarle Street. And he was extremely active there, both as a researcher and as a popularizer of science among young people, country-wide. At Southampton, he had built a well-funded group studying photophysics and photochemistry.

Phillips also studied the fluorescent properties and electronic energy migration in synthetic polymers, and he developed sensitive fluorescent probes in biological membranes.

A notable achievement in his research was the study of the photodynamic therapy of cancer in collaboration with University College Hospital, where his collaborator, Professor Stephen Bown, was based. Phthalocyanines[22][23] and porphyrins were regularly used as photosensitizers.

At one of his Christmas Lectures, broadcast by the BBC, given jointly with me on the topic of *'Crystals and Lasers'*, in 1987–8, Phillips demonstrated the efficacy of his and Bown's joint work on the photodynamic therapy of cancer.

Phillips' work at the RI, with more than a dozen research collaborators, was highly regarded worldwide. It was no surprise that Imperial College offered him a Professorship in their Department of Chemistry in 1989. He later became Head of that Department, and later still he became President of the Royal Society of Chemistry.

David Phillips did stalwart work in his lecture-demonstrations to schoolchildren. He also performed frequently abroad—in Singapore, Tokyo, and Moscow, where he studied after completing his Ph.D. He also served with distinction as Acting Director of the RI for one year prior to my taking over in 1986.

Another measure of his success in advancing knowledge is reflected in the numerous high-quality scientists who worked under his aegis. Twenty-one of them now hold senior posts (professorships or readerships) in the UK, France, Ireland, Switzerland, Australia, the US, Hungary, Japan, Mauritius, and Israel; and twenty other former students or postdoctoral scientists of his are employed as senior members in industrial and governmental organizations.

12.13.2 Richard Catlow

When David Phillips left the RI in 1989, I had little hesitation in recommending to the Council the person that I wanted as David Phillips' replacement: Professor Richard Catlow, who then held a joint appointment as Professor of Chemistry at the University of Keele and the Science Research Council's Synchrotron Laboratory in Daresbury, near Warrington.

As early as the 1970s, I had heard great praise for Richard Catlow's abilities from friends at University College London. They were overjoyed in the Department of Chemistry when they appointed him from Oxford, where he had distinguished himself both as a fertile researcher and imaginative teacher. In the intervening years, he and I interacted frequently, not only as reciprocal Ph.D. examiners for our respective research students, as we also carried out a piece of joint work, published in *Nature*,[24] that rationalized the seemingly rabbinical complexity of inter-growths in pyroxenic silicates.

What Richard Catlow had acquired, ahead of almost all other solid-state scientists in Europe, was a mastery of computer simulation, and a great awareness of what future developments there would be in both the speed and power of new generations of electronic computers and also the rapidly expanding methods of computation that were urgently needed by individuals, such as myself, who were concerned with elucidating the role of structures (and the feasible defects within them) in governing the catalytic and other properties of a wide range of solids. Richard Catlow had also mastered the use of centralized facilities, such as the synchrotron in Daresbury, and the neutron source in Grenoble. Richard and I, with a French postdoctoral, Dr Eric Dooryhee, of the University of Paris, were able to do pioneering studies on metal-substituted zeolitic catalysts, as well as locating templates in zeolites, and he, with Dr Clegg, of University College London, led the world in the field of microcrystalline diffraction.

Like David Phillips, when George Porter recruited him, Richard Catlow had built up an impressive team of graduate students and postdoctoral researchers at Keele and Daresbury. By the time I approached him in early 1990, he had already acquired an enviable international reputation.[25]

From the moment he arrived in Albemarle Street, he and I started some joint work, which continued for the next eighteen years. But we also carried out different distinct studies in the wide field of solid-state chemistry and, in particular, in the design and characterization of new heterogeneous catalysts, especially those that were environmentally benign, and capable of functioning in reactions that generated zero waste. A major part of our studies involved nanoporous solids (especially zeolites), which overlapped greatly with my studies of single-site heterogeneous catalysts that I had pioneered.[26,27]

With his extremely able colleagues, he became a world figure in his selected area of solid-state chemistry; and, as a consequence, he won several national and international prizes.

Again like David Phillips, he put in enormous efforts in popularizing science to young people, and even extended previous practice at the RI by reaching out to primary schoolchildren, not only in London and its environs, but also nationwide. And he, like David Phillips, was an admirable Deputy Director in his attention to the work required to sustain the RI and the DFRL (of which he later became Director in the period 2001–07) to achieve the fundamental goals of the RI.

A further advantage in having Richard Catlow at the RI was that he was an expert in the deployment of synchrotron radiation, as well as neutron beams, techniques that I and my associates relied on in the *in situ* study of catalysts.

Many of the postdoctoral scientists who worked alongside Richard Catlow and myself ((Drs Sankar and Raja (India), Maschmeyer (Australia), Bell (University College London), Cora (Italy), Rey (Spain), Wright (St Andrews), Zhou (China), Tennakoon (Sri Lanka), and Pickering and Williams (Cambridge)) now hold senior posts in academia in the UK and many other countries. Together, in the period 1990–2006, the Catlow—Thomas research teams published in excess of eight-hundred-and-fifty scientific papers that have, by now, been cited more than forty-thousand times by other workers engaged in the advancement of knowledge. Nearly forty of our former co-workers at the RI now hold senior academic posts in the US, Australia, India, Sri Lanka, China, Japan, Canada, Catalonia, Poland, Sweden, Italy, Russia, Taiwan, Germany, and Spain—ten are professors in this country, twelve occupy senior posts in industry (in the UK and abroad), and fourteen are in governmental laboratories or are civil servants. One of the professors in Australia, Thomas Maschmeyer, Head of Catalysis and Sustainability at the University of Sydney, was elected a Fellow of the Australian Academy of Sciences a few years ago. He has recently won the Australian Prime Minister Prize for Science 2020.

In regard to my collaboration with the Catlow group, it had been my intention—and it was approved by the RI Council when I had to retire as Director in 1991 (because of my wife's terminal illness)—that I would continue to pursue research, without remuneration, in the DFRL until I felt that I could no longer pursue further original, or other related, work. This was the way in which Sir Lawrence Bragg had pursued his work in George Porter's days as Director. I continued to do so until 2006. Thereafter less research was carried out at the RI. In due course, the workshop was closed, technicians were no longer employed, and the space was used for purposes other than research.

12.14   Discourse Entertainment

Such is the appeal of the RI that when invitations were sent, even to busy and eminent people, to come and dine with the Discourse speaker prior to the Discourse itself, seldom were there any refusals. Thus, senior political figures such a Lord Callaghan, Lord Cledwyn, Lord Elwyn-Jones (the former Lord Chancellor), Lord Wolfson, and Lord Lewis, and justices such as Lord Edmund Davies and Lord Hooson, were often present prior to and after the Discourses were presented. But my wife and I tried hard to invite to our flat many members of the RI, as well as other dignitaries.

Immediately after a Discourse ended, members and guests viewed the exhibition that the Librarian and Archivist, Irene McCabe, and the Discourse speaker had set up in the library, and refreshments were served to all attendees by the staff of the RI. There was usually a quartet playing in the background. One Friday evening, my wife introduced me to two distinguished elderly ladies who had come to attend a musically flavoured Discourse: Mrs Vaughan Williams and her friend, Lady Barbirolli.

Following a long-established practice—started, I was told, by the sociable Lady Bragg (wife or Sir Lawrence)—at about 10.30pm, the Director and his wife invited about twenty to thirty of the attendees to join them and the speaker and his or her guests to partake of further (liquid) refreshments in the Director's flat on a higher floor of the building. Conversation over drinks would continue beyond midnight. Very many interesting members of the RI joined us on such occasions. It gave me huge pleasure that they came from all walks of life – leaders in their fields of science, business, medicine, theatre, law, journalism and fashion. This facet of the Director's duties was especially exacting on my wife since preparations for Friday Evening Discourses commenced in the early afternoon and required meticulous planning. When, however, I look back at those late nights they were highly rewarding encounters that enriched our cultural life enormously.

It was my friend, Lord (Emlyn) Hoosen, who suggested to me that Sir Bernard Ashley, husband of Laura, should be invited to a Discourse dinner. In the brief conversation that evening, before he flew back by helicopter to Brussels, Sir Bernard, impressed by what the RI stood for, and especially to its commitment to young people, promised to send me financial support (which soon arrived) to set up the 'Laura Ashley Teenager Fellowship' to enable first-year university students to come and work for brief periods alongside one of the DFRL researchers during the summer vacation. This scheme worked admirably, and several research papers were published involving Laura Ashley fellows as co-authors.

12.14.1   Sam Wanamaker, Jonathan Miller, and Oliver Sacks

Whereas Sam Wanamaker came to the RI for Friday Evening Discourses at least twenty times during my period as Director, polymath Jonathan Miller came only once, to deliver a riveting Discourse on the importance of the live theatre.

Figure 12.16 *(a) Sam Wanamaker, American actor and producer, and the man chiefly responsible for the resurrection of Shakespeare's Globe Theatre. (b) Dr Jonathan Miller, opera and theatre director, writer, humourist, actor, and medical doctor. (c) Oliver Sacks, physician, best-selling author, and professor of neurology at the New York University School of Medicine. (Wikipedia open access)*

Sam Wanamaker, the American-born actor and director, who is credited as the person most responsible for saving the Rose Theatre and resurrecting Shakespeare's Globe Theatre, relished mingling with members of the RI on the theatrical occasions that were pursued every Friday. He was a charming and most engaging individual. I had several conversations with him, but I regret now not asking him what it was like acting opposite Ingrid Bergman on Broadway in 1942, and also, as Iago, opposite Paul Robeson's Othello, in Stratford-upon-Avon, in 1959.

Wanamaker stayed in London from 1952 onward, during the depths of the McCarthy era in the US, for he knew he would be blacklisted in Hollywood because he, like Robeson, had some sympathies with the Soviet Union, where they saw no signs of racial discrimination.

During the entertainment after Jonathan Miller's Discourse, we had a very interesting discussion about great actors. It surprised me when he said that the most impressive Shakespearean actor he had ever encountered was a Welshman who was at Cambridge with him as students. He then recited a speech from Shakespeare—as I recall it was *'Tomorrow and tomorrow and tomorrow, creeps in this petty pace from day to day...'*

When he recited this, I could detect the Welsh accent of someone from the part of South Wales where I was brought up. So I pressed him to tell me the name of this Welsh actor. He said that he had completely forgotten, and he had not been in touch with him since they both had left Cambridge. My curiosity intensified, and I told him that Wales is so small, and the Welsh-speaking community smaller still, that if he could remember this Welsh actor's name, I would surely be able to find him. No further response came from Miller.

Several months later, it dawned on me that Miller was playing a game. When he recited the passage from Shakespeare, he was impersonating me! I never had the opportunity of seeing him face-to-face again.

Oliver Sacks never lectured at the RI, but he was—as he told me—in love with the place, for as a teenager he accompanied his uncle (Dave—of *'Uncle Tungsten'* fame) to the library and basement of the RI, mainly to acquire more knowledge (as described in Chapter 11) about the extraordinary Humphry Davy.

As a child in North London, he became a close friend of Jonathan Miller, a friendship that lasted his whole lifetime. They both attended Westminster School, which they each greatly enjoyed. And even when Miller went to Cambridge for his undergraduate studies and he to Oxford for his, they still kept in close touch with one another. They later frequently met in New York, where they each pursued their careers.

My friendship with Oliver Sacks started in 2002, after he came to visit me at Peterhouse, Cambridge, to find out, as he put it, what kind of person I was. He had read my book review, published in *Angewandte Chemie Intl Ed.*, of his memorable *'Uncle Tungsten'*, which I had said was one of the most enjoyable I had ever read. I entitled it *'An Omnivorous Curiosity, a Sense of Wonder, and a Taste for the Spectacular'*.

Thereafter, we began to exchange a few letters with one another, and for the last dozen or so years of his life they became more regular. He came to listen to my lecture *'On the Genius of Michael Faraday'*, at New York University in May 2014; and he entertained my wife and me at his home in Horatio Street, afterwards. He told me that he had noticed that before my lecture, for twenty minutes or so, I was not seen in the lecture theatre. He had read what Sir Lawrence Bragg had once said, namely that to expect a lecturer to enter into social conversation and small talk before embarking on an important lecture was the pinnacle of inconsideration. It was Bragg's practice, and I had adopted it, to secrete oneself in some invisible corner until a few minutes before one was called to lecture and be introduced. Faraday had similar views, which is why, prior to any Discourse given at the RI, the Discourse speaker is led into a quiet room to collect his thoughts before performing.

In Chapter 11, we saw how the young Sacks became aware (in the RI) of Davy's discovery of the alkali metals: sodium, potassium, rubidium, caesium, etc., and how remarkably these metals behaved in liquid ammonia.

To illustrate Oliver Sacks's brilliant gift of communication and mastery of metaphor, it is relevant to quote a brief passage from his posthumously published book *'Gratitude'*, in which he wrote: *'And now, at this juncture, where death is no longer an abstract concept, but a presence—an all-too-close, not-to-be-denied presence—I am again surrounding myself, as I did when I was a boy, with metals and minerals, little emblems of eternity.'*

As mentioned in Chapter 1, the periodic table enchanted him, and throughout his life he made pleasant reference to it. He never ceased to wallow in the joys that could be derived from the periodic table. As a child in London, from the age of ten upward, its elements became his companion: *'Times of stress throughout my life had led me to turn, or return, to the physical sciences, a world where there is no life, but also no death.'* When my wife and I sat for a few hours in his book-lined study in New York in 2014, he showed us a range of beautiful specimens of some of the most attractive elements (gold, bismuth, copper, etc.) of the periodic table. In a charmingly illustrated small box, sent to him by a so-described *'element-friend'* in

England, the label on it read *'Happy Thallium Birthday'*, a souvenir of his eighty-first birthday (in July 2014). The atomic number of thallium, Tl, is 81.

Towards the end of his life, Oliver Sacks urged me to write combined biographies of Davy, Faraday, Tyndall, Dewar, Rayleigh, and the Braggs. I replied that I was thinking of doing something like that, and I asked him if he would be prepared to write the Foreword to it, to which he agreed with alacrity and delight.

The unique and varied qualities, abilities, and skills of Oliver Sacks are known to countless people who have read his numerous fascinating books and his evocative essays. But, as someone I got to know in person, away from the letters that passed between us, he was an utterly unforgettable human being, blessed with so many of the qualities that one wishes to see in others and in oneself. I shall treasure over the years to come the memory of him for as long as I live; and I shall periodically consult those carefully crafted letters he sent to me.

12.15   Carl Sagan

It is invidious to single out one Discourse for special mention in a monograph of this kind. Before I became Director, and also afterwards, I witnessed several outstanding Discourses and Christmas Lectures.

However, there is one Discourse, which I know was greatly admired by my predecessor, George Porter, and it was one that I myself witnessed, namely Dr Carl Sagan's account of the imminent arrival of Halley's Comet, given in October 1985.

George Porter had earlier drawn my attention to the extraordinary ease with which Sagan had presented his Christmas Lectures on the *'Planets'* in 1977. Most Christmas Lecturers take a great deal of time to set up all the demonstrations with the various assistants, such as Bill Coates and Bryson Gore. (I put great strain on my marriage when I spent so much time preparing all the demonstrations for my Christmas Lectures on *'Crystals'* in 1987.) But George Porter alerted me to the brilliant ease with which Carl Sagan improvized gadgets and he succeeded to present a superb lecture when he spoke *'On the Eve of the Comet'* in October 1985. I was present in the audience on that evening and I was immensely impressed, not only by Sagan's scintillating Discourse, but by the engaging and fluent summary that he provided for the RI programme of events that autumn. So impressive do I consider Sagan's summary that I repeat it here. It represents a fine example of what RI speakers can achieve when they attain their best as communicators of knowledge.

'The 1985/86 return of Halley's Comet is the 30th consecutive apparition of this comet in human chronicles—although it was probably captured into the planetary part of the solar system during the last ice age. Comets ultimately arise from a vast cloud of 100 million million icy worlds, each typically the size of a city centre, a few of which are, on occasion, gravitationally perturbed by a nearby star or interstellar cloud and induced to enter the inner solar system where they become visible to Earthbound astronomers. Comets seem to be debris left over from the formation of the solar system. It is conceivable that a significant fraction of the oceans of the Earth and the organic matter that led to the origin of life were

Figure 12.17 *Dr Carl Sagan.*
(Wikipedia open access)

brought by comets, and that through episodic or periodic impacts of comets with the Earth, massive biological extinctions have been produced; the asteroid or comet that impacted the Earth 65 million years ago seems to have extinguished the dinosaurs and made possible the success of our mammalian ancestors. Far from their traditional roles as omens, comets might more appropriately be considered a continuing benign influence on the human species. For the first time in its hundreds of perihelion passages, Halley's Comet will in this apparition be greeted by a flotilla of spacecraft representing 20 nations, which should dramatically improve our knowledge of these enigmatic small worlds.'

12.16   Kirill Zamaraev

Kirill Illyich Zamaraev, a Union of Pure and Applied Chemistry (IUPAC, 1993–5), holder of the Centenary Medal of the Royal Society of Chemistry (1995), and former Director of the Boreskov Institute of Catalysis of the Siberian Branch of the Russian Academy of Sciences, died in the prime of his brilliant scientific career. News of his passing cast gloom over the twelve-hundred delegates at the Eleventh International Congress on Catalysis, from 1–7 July 1995 at Baltimore, where he was scheduled to deliver one of the plenary lectures on *'Photocatalysis: State of the Art and Perspectives'*.

Kirill Zamaraev was an extraordinary individual. Equally adept as a theoretician and experimentalist, he was an exceptionally versatile chemical physicist, who communicated the gifts and insights of his teachers in Moscow (among them Lev Landau and Peter Kapitza) to a large family of chemists and engineers. Widely read in several languages, he had a finely developed taste for literature, the theatre, and ballet. He was a born leader and astute diplomat who instilled confidence into others and inspired great efforts from his colleagues. He was also a life-enhancing soul, whose company and friendship was valued by all who met him.

During the upheavals and changes consequent upon the disintegration of the Soviet Union, Zamaraev played a leading role in the transformation of Russian science and its adoption of the market-force economy. In particular, he assisted in the establishment of a network of Federal Research Centres throughout Russia. All this occurred when he held office as President of the IUPAC, which itself demanded an extensive and punishing round of worldwide visits to international conferences and workshops, as well as constant liaison with the IUPAC secretariat in England, at Oxford.

In 1974–5, as part of a US–Soviet exchange programme, Zamaraev held a series of visiting professorships at the universities of Cornell, Stanford, and Chicago, where his superb fluency in English and the authoritative flair of his scientific understanding were to impress his American hosts. In 1977, he was selected as the heir apparent to Academician Boreskov as head of the largest institute of catalysis in the world (employing more than a thousand people), set up as part of the Krushchev experiment, in Akademgorod, Novosibrisk, in Siberia. He took a large team of bright Muscovite chemical physicists with him, thereby broadening the horizons of the already formidable Boreskov Institute so as to encompass fundamental and applied studies ranging from *ab initio* quantum mechanics to enzymatic chemistry. In 1984, he took over as Director, a post that he held until 1992, when he decided to relinquish some of his administrative duties in favour of the more active pursuit of research.

By 1987, Zamaraev had become a full member of the Academy of Sciences of the Soviet Union, and was in great demand worldwide as a plenary speaker on a host of topics ranging from industrial applications, chemical engineering practices, and laboratory 'model' studies of catalysis.

When the Royal Society introduced its Kapitza Fellowship Scheme shortly after the collapse of the Soviet Union, Zamaraev was one of the first to be appointed (for six weeks based in the DFLR, but with visits to the universities of Cambridge and Wales at Cardiff). Earlier, in 1988, his group at Novosibirsk had started a collaboration, which continued up to his death, with my group at the RI on the catalytic properties of zeolites. Our prime aim was to understand how the catalytic breakdown of certain molecules is influenced by confinement within the pores and cavities of zeolitic solids. As one of my postdoctoral workers at the RI, Dr Carol Williams, spoke fluent Russian, we arranged to send her as an investigator to his Institute in Novosibirsk. She also spent some of her two- year stay in the USSR with Academician Kazanzky in The Institute of Organic Chemistry, Moscow. Together, we aimed to understand how the catalytic breakdown of certain molecules is influenced by confinement within the pores and cavities of micro- and meso- porous hosts.

Figure 12.18 *Dr Carol Williams, who played a crucial role in the Zamaraev—Thomas collaboration.*
(Courtesy Dr Carol Williams)

A particularly interesting research paper emerged from Carol Williams' joint work with me and Kirill Zamaraev.

Zamaraev's memorable Centenary Lecture, given in the theatre of the RI, on 25 January 1995, still rings in my ears. In it, he described *inter alia* how at the Boreskov Institute they had: (i) sweetened natural gas (i.e., succeeded in eliminating the hydrogen sulphide from methane); (ii) harnessed the Sun's rays to remove the same gas photocatalytically from naturally contaminated inland seas in Russia; (iii) used immobilized enzymatic catalysts to convert natural gas to methyl alcohol; (iv) designed a catalyst to convert methyl alcohol to formaldehyde; (v) devised a means of converting wasteful sulphurous by-products from industrial plants into sulphuric acid; and (vi) set up in the Ukraine a thermocatalytic converter that stored solar energy chemically and released it by the ingenious use of catalysts.

All this and more is described in one of his last major articles (published in *Topics in Catalysis*).

While in a Moscow hospital in the autumn of 1994, Zamaraev wrote a beautiful booklet that chronicles the achievements of Russian scientists in catalysis from the days of Mikhail Lomonosov in the eighteenth century onward. It contains a wealth of evocative entries, including the fact that composer and chemist Alexander Borodin described the so-called aldol condensation ($2CH_3CHO \rightarrow CH_3CH(OH)CH_2CHO$) in 1872.

When Kirill Zamaraev visited Cambridge in the summer of 1993 with his wife, Mila, he quoted his beloved Alexander Pushkin: '*How many and marvellous are the discoveries prepared for us by the spirit of enlightenment, by experiment, the child of error and effort, by genius, the friend of paradox, and by that divine inventor, chance.*'

I often reflect that it was as a result of chance that he and I met, at the Eighth Congress on Catalysis in Berlin, in July 1984.

12.17   Photographs and Portraits of Some Other Notable RI Performers

A minute fraction of the notable workers and lecturers who have performed at the RI is shown in Figures 12.19–12.21 to further reinforce the unique character of its cultural contribution.

In his later years as Master of Trinity, 'J. J. Thompson' behaved in an increasingly embarrassing and eccentric way. Sir Alan Hodgkin, another famous Discourse speaker, and later Master of Trinity, recalls in his autobiography

(a)   (b)   (c)   (d)

(e)   (f)   (g)   (h)

Figure 12.19 *(a) Charles Babbage, the pioneer of modern programmable computing; (b) Nikola Tesla, the remarkable Serbo-American champion of alternating current; (c) Kathleen Kenyon, the archaeologist whose work revealed, inter alia, that the city of Jericho is arguably the oldest continuously inhabited one in the world. (d) Roald Hoffmann, a poet and dramatist, educator, and popularizer of science; (e) Mary Archer, who started her important work on solar energy at the RI. (f) Francis Crick, Fullerian Professor of Physiology at the RI, who gave a spectacular Discourse in 1976 on 'The Gene', (g) Sir William Crookes, a major figure at the RI, where he interacted with Faraday, Rayleigh, and many others; and (h) Frederick Sanger, who gave a Discourse in 1978 on the DNA sequence of an organism, is the only British scientist to have been awarded the Nobel Prize twice.*

(Courtesy (a), (b), (c), (d) Wikipedia open access, (e) Mary Archer, (f) Laboratory of Molecular Biology, Cambridge, (g) RI (h) Laboratory of Molecular Biology, Cambridge)

(a) (b) (c)

(d) (e) (f)

Figure 12.20 *Other notable performers at the RI: (a) Erwin Schrödinger, a founding father of quantum mechanics; (b) Dorothy Hodgkin, the first British woman to win the Nobel Prize (for her elucidation of the structures of penicillin, cholesterol, and vitamin B_{12}; (c) John Maynard Keynes, the foremost economist of his age; (d) Yehudi Menuhin, one of the world's greatest violinists; (e) H. G. Wells, now known as the father of science fiction; and (f) Hermann von Helmholtz, the eminent German physiologist, psychologist, physicist, and philosopher.*
(Wikipedia open access, public domain)

'*Chance and Design*' an amusing story relating to 'J. J.': '*On seeing that W. B. Yeats was to be sitting next to him in Hall, JJ opened the conversation with the remark "Been writing much poetry lately, Mr Keats".*'[28]

Other giants, such as Alfred Russel Wallace, co-discoverer of the theory of evolution, Frederick Soddy, who coined the word 'isotope', and other notable individuals such as the actor Henry Irving, the cosmologist A. S. Eddington, and Patrick Blackett, a pioneer of cosmic ray science, all gave discourses at the RI. J. J. Thomson gave some fifteen and Ernest Rutherford more than a

Figure 12.21 *Image of two Nobel Prizewinning Cavendish physicists. At left, Emeritus Professor J. J. Thomson, who supervized Ernest Rutherford (right). There was much rivalry between them. Here they are seen preparing for the annual departmental photograph a year or so before Rutherford's untimely death in 1937.*
(Courtesy and copyright Cavendish Laboratory, University of Cambridge)

dozen, his last on the transformation of heavy elements; only a few months before his death. Rutherford always showed spectacular demonstrations in his Discourses.

This collection of names and photographs is but a mere fraction of the galaxy of eminent individuals who have presented their work at the RI.

I have omitted to describe the short, but very active tenure of Sir Eric Rideal, who occupied the Fullerian Professorship of Chemistry for a brief period (1946–9). Figure 12.22 is a characteristic sketch of him (with the Gibbs adsorption isotherm in the background) when he used to stroll daily through the laboratories of the DFRL. He left the famous Colloid Chemistry Laboratory,

Figure 12.22 *Sketch (artist unknown) depicting Sir Eric Rideal on his daily peregrinations through the laboratories of the DFRL.*
(Courtesy Professor B. A. Pethica, Princeton University)

University of Cambridge—one of the finest in the world—to take up his RI post. But he felt it necessary to resign from the RI owing to his wife's ill health. While at the RI, Rideal initiated the continuing series of *'Advances in Catalysis'*.

It reminds all those who enter its precincts how remarkable an institution Rumford and his supporters created in a corner of Mayfair, London, in 1799.

REFERENCES

1. G. Porter, *New Scientist*, **1977**, September, 802.
2. I can still hear George Porter saying: *'After all, the* Proceedings of the Royal Institution *was started by Faraday and they constitute a unique record of the advancement of science and technology.'*
3. It is alleged that in 1846 there occurred an event of both mythopoeic and scientific significance. Charles Wheatstone was due to give a Discourse at the RI on his electro-

magnetic chronoscope. He was incurably shy and panicked and at the last moment he left the building hurriedly. Faraday had to present a Discourse in his place.[5]

4. K. Lonsdale, *'Crystals and X-Rays'*, G. Bell & Sons, London, **1948**.
5. My Ph.D. thesis was entitled: *'The Significance of Structure in Carbon-Gas Reactions'*.
6. Z. Ludmer, M. D. Cohen, J. M. Thomas, and J. O. Williams, *Proc. Roy. Soc. A*, **1971**, *324*, 439.
7. M. Glazer and Patience Thomson (eds), *'Crystal Clear: The Autobiography of Sir Lawrence and Lady Bragg'*, Oxford University Press, **2015**.
8. M. Glazer, a stalwart supporter of the RI, who attended Sir Lawrence Bragg's lectures when he was a schoolboy, and who has himself given Discourses at the RI. (See Afterword, below.)
9. W. H. Thompson, *Trans. Roy. Soc. Edin.*, **1869**, *34*, 15.
10. Topology has recently been found to be highly relevant, by the Cambridge chemist Professor Sophie Jackson, in the structures of so-called alpha-helices in protein structures (see Figure 4).
11. E. Witten, *Proc. Int. Congr. of Mathematical Physics*, **1988**, July.
12. S. E. Jackson, A. Susma, and C. Mitchelatti, *Current Opinion in Struct. Biology*, **2017**, *42*, 6.
13. The Frisch–Pierls memorandum was the document based on the calculations of Rudolph Pierls and Otto Frisch made in the University of Birmingham.
14. A very distinguished Indian physicist who worked with Kathleen Lonsdale in the DFRL in the early 1930s.
15. J. M. Thomas, *'Michael Faraday and the Royal Institution: The Genius of Man and Place'*, IOP Publishing, Adam Hilger (now published by Taylor & Francis), **1991**, p. 72.
16. M. S. Longair and J. Waldron, 'Sir Alfred Brian Pippard, *Biographic Memoirs of Fellows of the Royal Society*, **2009**, *55*, 201.
17. Professor Humphreys (now Sir Colin) demonstrated in his Discourse that the Star of Bethlehem was a comet in 5 BC and that the date of the birth of Jesus Christ was in the period 9 March to 4 May. (See C. J. Humphreys and W. G. Waddington, *Nature*, **1983**, *306*, 743. Later work by these two authors established that a solar eclipse occurred in 1207 BC—see *A and G*, **2007**, *58*, 5.)
18. F. Gottfried, *Proc. Roy. Inst.*, **1990**, *62*, 127.
19. Later, BBC viewers were to see Dr Neil MacGregor often perform such lectures.
20. He shared it in Chemistry in 1981 with the Japanese theoretical chemist Professor Kenichi Fukui, who later became an Honorary Member of the RI.
21. J. M. Thomas, 'Ahmed Hassan Zewail', *Biographical Memoirs of Fellows of the Royal Society*, **2020**, *68*. 431.
22. It was at the DFRL that the structure of phthalocyanines were determined by a contemporary of Kathleen Lonsdale, J. Monteith Robertson, who became one of the world's leading crystallographers.
23. J. M. Robertson, 'X-Ray Analysis and Applications of Fourier Series Methods in Molecular Structures', *Rep. Progr. Physics*, **1937**, *4*, 332.
24. R. A. Catlow, J. M. Thomas, S. Parker, and D. A. Jefferson, *Nature*, **1982**, *295*, 658.
25. One of the groups that he had pursued joint research work with was based in the East Berlin Academy in Alderhof. In particular, he was interacting with Dr Joachim Sauer and his wife, the future Chancellor of united Germany, Angela Merkel, whom I first met in the Director's flat.

26. J. M. Thomas, *'Design and Application of Single-Site Heterogeneous Catalysts: Contributions to Clean Technology, Green Chemistry and Sustainability'*, Imperial College Press, London, **2012**.
27. K. D. M. Harris (ed.), *'Selected Papers of Sir John Meurig Thomas'*, World Scientific Publishers, Singapore, **2017**.
28. A.Hodgkin, *'Chance and Design: Reminiscences of Science in Peace and War'*, Cambridge University Press, **1994**.

Afterword

Over the years, many individuals of differing expertise and different national-ities have displayed profound affection for the Royal Institution (RI). When they explain why, some of them recall the halcyon days of Davy and Faraday, who made the RI a much more widely cultured centre than, for example, the Royal Society at the time. Others, such as an early President of the US National Academy of Sciences, Joseph Henry, admired it for the extraordinary number and range of discoveries that emerged from it. Yet others recall with excitement the memorable lecture-demonstrations to schoolchildren, pioneered by Faraday in his Christmas Lectures and brilliantly extended by Sir Lawrence Bragg more than a century later.

Moreover, there is a large body of crystallographers who regard the RI as being, for several decades, one of the most powerful centres of X-ray crystal analyses in the world. Equally, molecular biologists associate it as the birthplace of their sub-ject, and in particular the place where, later, the structure of an enzyme was first determined.

The very building and its laboratories, library, museum, and charming lec-ture theatre (Figure A.1) make it a great attraction to scientists and non-scientists alike.

In closing, I have, included as an epilogue a few specific and favourite inci-dents that are of special interest to me, that reinforce the uniqueness of the RI, and which illustrate the esteem in which the RI is held. I also reflect briefly on the part it played in my research, and where it sits in the present-day scientific landscape.

A.1   Max Perutz's Eightieth Birthday Symposium

In 1994, when Max Perutz reached the age of eighty, his colleagues at the Laboratory of Molecular Biology, in Cambridge, arranged a one-day sympo-sium in his honour in the lecture theatre of the RI. By that time, Perutz had given some ten RI Discourses, and he and his colleague John Kendrew had beenVisiting Readers in the Davy–Faraday Research Laboratory (DFRL) from 1953–66, having been appointed by their mentor, Sir Lawrence Bragg, whom they revered.

But Max Perutz also possessed a finely tuned historical sense—one had only to raise the topic of Vienna, his home city, to be convinced of this—so it was not surprising that he requested that his eightieth milestone should be celebrated at

Figure A.1 *The famous Royal Institution (RI) lecture theatre, built in 1801 to hold nine-hundred people. Because of the Safety at Work Act 1972, it now holds about five-hundred. Faraday, as an apprentice, when he listened to Davy's lecture in 1812, is thought to have sat behind the clock. The inset shows Max Perutz lecturing at his eightieth Birthday Symposium at the RI (see text) in 1994.*
(Courtesy Shell Education Supplement and Sir Alan Fersht)

the RI, which had played such an important part in his scientific life. Furthermore, the historicity of the place added an extra appeal.

Figure A.2 shows Max, sandwiched between Aaron Klug and David Blow and others, standing outside the lecture theatre at the RI and in front of a famous painting that pictures a Discourse by Sir James Dewar (on liquid hydrogen) in early 1908.

Max was keen to convey to the delegates from overseas the dramatic atmosphere that pervaded Discourses at the RI. Eminent scientists who had come to the Perutz celebration from as far afield as San Francisco, Chicago, Harvard, the US National Institutes of Health, several European centres, and Japan, were impressed by the atmosphere engendered in the RI lecture theatre.

Figure A.2 *A view outside the entrance to the lecture theatre of the RI taken on the occasion of the eightieth birthday celebration of Max Perutz (centre). To his immediate right is Sir Aaron Klug and next to him is Michael Rossmann, and on his right are, first, Wim Hol and then Tony Kossiakoff. On Perutz's left is David Blow, followed by Peter Colman, Don Wiley, and Wayne Hendrickson. Behind the group is a painting of a typical Friday Evening Discourse audience in the early 1900s, at which Sir James Dewar demonstrated the properties of liquid hydrogen. In Dewar's audience are Lords Rayleigh and Kelvin, A. J. Balfour (the Prime Minister), and G. Marconi.*
(By kind permission of Dr Richard Henderson)

A.2   Schools and Christmas Lectures

In my discussions over the years with university students in Oxford, Cambridge, and the constituent colleges of the University of London, I have often been struck by how many of them were prompted to pursue science subjects at university partly because of the magnificent schools lectures presented by Sir Lawrence Bragg from about 1956–66, and partly because of the riveting Christmas Lectures presented at the RI for the children living in the environs of London. Numerous impressive Christmas Lectures were mentioned to me as vivid memories; for example, David Attenborough's 'Language of Animals' and those by Carl Sagan, Colin Blakemore, Richard Dawkins, Malcolm Longair, and Nancy Rothwell (see Figure A.3).

Figure A.3 *(a) David Attenborough (with a ring-tailed lemur); (b) Richard Dawkins, whose topic was* 'Growing up in the Universe'; *(c) Nancy Rothwell (*'Staying Alive'*); and (d) Colin Blackmore (*'Common Sense'*). All of these presented successful Christmas Lectures.*
(Courtesy (a) BBC photo library, (b) Public domain (c) University of Manchester, (d) Colin Blackmore)

But it is not only university students who often wax eloquent about attending schools lectures at the RI. Very many university professors also state that their education owes a great deal to the mesmeric talks that Sir Lawrence Bragg and others gave at the RI when they were of school age. For example, Professor A. M. Glazer (Clarendon Laboratory, University of Oxford) recalls vividly the occasion in 1958 when he attended a schools lecture given by Sir Lawrence Bragg (Figure A.4).

A.3   The Enzyme Structure Breakthrough

In earlier sections of the book (see Figure 1.7 of Chapter 1, and Chapter 7), the contributions made to molecular biology at the RI have been outlined. Both W. T. Astbury and the polymathic J. D. Bernal (Figure A.5) were the progenitors of such work.

Figure A.4 *Professor A. M. Glazer, as a teenager, is seen in the top left corner of the upper part of this dual photograph, Sir Lawrence Bragg occupying the lower half.*
(Courtesy Professor A. M. Glazer)

Figure A.5 *J. D. 'Sage' Bernal.* Max Perutz, who studied under him, remarked: 'Knowledge poured from him as from a fountain, unselfconsciously, vividly, without showing off, on any subject under the sun....He was the most incredible, magnetic and interesting character I'd ever met. In my native Austria, I never knew that anyone like that ever existed.'
(Courtesy Professor Robin Perutz)

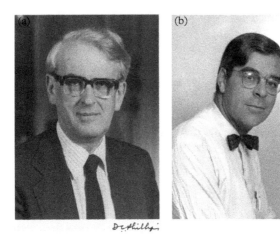

D. Phillips

Figure A.6 *(a) Professor David Phillips; (b) Professor Uli Arndt.*
(Courtesy (a) the Royal Society, (b) Laboratory of Molecular Biology, Cambridge)

However, the cognoscenti among molecular biologists and biophysicists point to the outstanding work of David Phillips (later Lord Phillips of Ellesmere), Uli Arndt, A. C. T. North, and Louise Johnson in their brilliant investigations into the structure and mode of operation of lysozyme, work acknowledged as being the beginning of structural enzymology, which is now pursued in laboratories worldwide.

Two of the scientists greatly influenced by Phillips and Arndt, Professor (later Dame) Louise Johnson—she was Phillips' Ph.D. student—and Professor Greg A. Petsko, of Harvard and Cornell—he was Phillips' postdoctoral collaborator—have testified[1][2] that the introduction of the linear diffractometer, designed and built at the RI by Phillips and Arndt (Figure A.6), led the way to a new era in crystal structure determination using X-rays. Prior to their breakthrough, all X-ray diffraction data were collected on film. The Phillips—Arndt instrument paved the way to fast detection and collection of X-ray scattering data, which led to an acceleration in progress throughout molecular biology and adjacent subjects.

A.4 Reflections on the Role as Director of the RI in my career

A.4.1 Leaving a Highly Effective Research School in Cambridge

In 1986, I was greatly honoured to be Head and 1920 Professor of Physical Chemistry at the University of Cambridge. Never in my wildest dreams when

I was appointed assistant lecturer at the University of North Wales in 1958 did I ever think I could be elected, in due course, to one of the UK's premier chairs in my subject. I still recall the thrill I had when Dame Rosemary Murray, the Vice-Chancellor of Cambridge, rang me in the spring of 1977 to offer me the Headship of Physical Chemistry. (I subsequently learned that it was one of her predecessor's, Professor J. W. Linnett, who, after spending two days in my Department at Aberystwyth in 1976, was prompted to tell his colleagues at Cambridge that he wanted me to be his second professor. Alas, Professor Linnett died suddenly, and I became his successor!)

After George Porter rang me in 1986 to ask if I was interested in following him at the RI, I became quite unsure. While it would probably be an even greater honour to be Director of the RI and, in so doing, to follow in the footsteps of Faraday and Davy at one of the world's most famous centres of research and popularization of science, on the other hand I had an exceptionally gifted team of researchers in my Department in Cambridge at the time. A glance at Figure A.7 is pertinent.

It conveys well the brainpower, potential, and achievements of my team, which focussed on a wide range of fundamental problems in the whole corpus of chemistry—physical, organic, inorganic, mineralogy, geochemistry, nanoscience, heterogeneous catalysis, and the development of a unique range of new techniques (now widely used in surface and materials science).

My research style was not only to solve key and interesting scientific problems, but also to develop fundamentally new techniques to do so. I was exceptionally well-equipped with some of the best available high-resolution electron microscopes, as well as solid-state nuclear magnetic resonance (NMR), electron energy loss spectroscopy, a rotating anode X-ray diffraction apparatus, and a conventional range of other spectroscopies—infrared, ultraviolet-visible, X-ray photoelectron, and sensitive adsorption apparatus.

Figure A.7 *The Solid-State Chemistry Group, at the Department of Physical Chemistry, University of Cambridge, in 1986.*
(Courtesy John Meurig Thomas)

In Figure A.7, I am sitting in the centre of the front row (six from the right) and to my right is Bill Jones, my former Ph.D. student at Aberystwyth. Such were his chemical and administrative qualities that he later became Head of the whole Department of Chemistry in Cambridge—a man of exceptional experimental skills. On his right is Professor Jacek Klinowski, who joined me on soft postdoctoral research money from Imperial College in 1979. A graduate in mathematics from the Jagiellonian University, Kracow, he became one of the world's leading solid-state NMR experts. Two places to his right is Dr Adrian Carpenter, now Reader in the Wolfson Brain-Imaging Centre in Cambridge. To my immediate left is David Jefferson, Reader in Chemistry and a leading electron microscopist. Next to him are G. R. Millward and B. G. Williams, very inventive electron microscopists, one an expert in Compton scattering. At the extreme end of that row is a Visiting Professor, Jack Lumsford from Texas A&M, an expert on selective oxidation catalysts. Immediately behind him in the second row is a Japanese Ramsay Scholar, Dr Wataru Ueda, who later became Director of Catalysis Research at Hakkaido University; and next to him is Dr Rik Brydson, now Head of Materials Engineering at the University of Leeds.

In the second row, the woman immediately behind Bill Jones is Carol Williams, who, being fluent in Russian, distinguished herself in the 1990s at the Boreskov Institute in Novosibirsk, in the Siberian Academy of Sciences, with whom I had started a collaborative programme with its Director, Academician Kirill Zamaraev. Two places to her left is Lynn (now Dame Lynn) Gladden, Fellow of the Royal Society, former Head of the Department of Chemical Engineering at Cambridge and former Pro-Vice-Chancellor of the University of Cambridge. Now she is Head and the Executive Chair of the UK's Engineering and Physical Sciences Research Council.

In the back row are some very bright scientists from Taiwan, China (Jilin), and Sri Lanka (D. T. B. Tennakoon, later Head of the University of Sri Lanka in Peradeniya). Notable also are Professor K. D. M. Harris, Distinguished Research Professor at Cardiff University, and next to him, visiting my group from the University of Edinburgh Professor Anthony Harrison, now Head of the National Diamond Light Source in Harwell. Next to him is Dr Andrew Nowak, of the Shell Company, and finally, Dr Subramanium Yashonath, Research Professor at the Indian Institute of Science, Bangalore.

A.4.2   Could I Emulate What George Porter Did at the RI?

When I discussed with my close friend Max Perutz whether I should become a candidate to follow George Porter, he advised against it: '*You will not be surrounded by a teeming collection of bright postgraduates and postdoctoral workers, as you are in Cambridge.*' Yet what Porter had accomplished at the RI as a research investigator was very impressive. He had collected a relatively large school of photochemists and physicists, as well as others—people such as Graham Fleming (later Fellow of the Royal Society), Geoffrey Bedard, Tony Harrisman, and Mary Archer. He had

Figure A.8 *The President of the Academia Ciencia, Professor Liu Jiaxi, being shown the electron microscopes in the Solid-State Chemistry Group, at the Department of Physical Chemistry, Cambridge. (Courtesy John Meurig Thomas)*

also ensured that he had assembled a fine collection of research technicians, along with a good workshop, a fine glass blower, an electronics expert, and the remarkable Bill Coates, the supremely able general factotum. Also, the number of research students in his team was significant, one of whom, Professor David Klug, is now at Imperial College. And he and Professor David Phillips, as his Professor of Natural Philosophy, shared many common facilities.

So, at first, I said I was not interested in applying for the Directorship. However, during an extended visit to many premier research centres in China (at universities in Beijing, Fudan, and Fujian), where I had been invited to lecture by the President of Academia Ciencia, Liu Jiaxi (who had earlier visited my lab in in Cambridge (see Figure A.8), my wife Margaret and I began to think afresh of the RI. I changed my mind, and was lucky enough to be appointed.

From the outset, I felt it was within me to do a decent job of the popularization of science. I quietly felt that insofar as living up to the lecturing skills of Davy and Faraday, I could ultimately make a go of it. But I pondered deeply whether I could run a viable research centre in my fields, which were very different from those pursued by Porter).

The thought of Miss Irene James, my physics mistress at my secondary school in Carmarthenshire, South Wales, kept recurring in my mind.

A.5   Miss Irene James

Miss Irene James (Figure A.9) was my physics mistress at the grammar school that I attended in the Gwendraeth Valley, in Carmarthenshire, South Wales, from 1944–51. She was an outstanding teacher, who not only ensured that we were conversant with all the necessary principles of the subject, she also would give us

Figure A.9 *Miss Irene James.*
(Courtesy John Meurig Thomas)

fascinating asides about individual scientists, e.g., that Newton was the son of an illiterate farmer, that Galileo ran into trouble with the Church, and that Rayleigh derived (and she showed us how) the inverse fourth power law for the scattering of light by 'dimensional analysis' (see Chapter 5).

One afternoon, when I was about fourteen, she started talking about Michael Faraday, whom she introduced as a blacksmith's son from London. She explained how Faraday had long pondered what the relationship was between electricity and magnetism and whether they could be interconverted.

She amplified further Faraday's unique ability as an experimentalist and as one of the greatest discoverers in the history of science. She was obviously in awe of Faraday. And she emphasized that he had scaled the great heights of world science, even though he had left school in his early teens to work in a bookbinder's shop, and that he knew no mathematics, except arithmetic.

She obviously kept abreast of current physics literature and talked to us about Patrick Blackett and his work on cosmic rays. In fact, a very bright boy, a year ahead of me, was guided by her to study at Manchester University and joined Blackett's research group.

Even afterwards, as an undergraduate in Swansea University, where I was well taught in chemistry and physics, whenever Faraday's name was mentioned, I thought of Miss James, and her admiration for him. In due course, Miss James, being the excellent educator that she was, left my grammar school to become

Vice-Principal of Carmarthen Training College, where she remained until her retirement, and where she was an important influence on its growth.

In the audience at the RI, on 8 November 1986, when I gave my inaugural Discourse on *'The Poetry of Science'*, Miss James was there, to my and her great pleasure.

A.6   Spending Time with Bill Coates

I quickly realized on taking up my duties at Albemarle Street that one of the great qualities that lecturer's assistant and prime technician Bill Coates possessed was that he knew the recipes and experimental procedures of essentially all the lecture-demonstrations that Faraday had carried out in his long career as an RI speaker. This was an invaluable experience, which enabled me in my later RI career to re-enact several of Faraday's famous lectures.

Bill Coates and I sat together for hours watching the films of the Christmas Lectures that Sir William Henry Bragg had presented from the 1920s. These were marvellously instructive. (And they had influenced countless young people in the environs of London, notably Dorothy Hodgkin.)

In my early days at the RI, I often doubted whether I could ever step into the shoes of Faraday (or Davy) as experimentalists.

So I concentrated in my early days at the RI on talking to and carrying out lecture-demonstrations with Bill Coates. On the very first night that I spent alone in the RI—my wife and children were still in Cambridge—Bill Coates told me a story that disturbed me. He said that several RI lecturers, including himself on occasion, felt the physical presence of Faraday next to them as they presented their lectures. Bill even said that many members of the RI believed that the ghost of Faraday roamed the building!

The Director's bedroom is situated immediately below a third floor DFRL laboratory. At about 2 am on my first night, I heard footsteps emanating from the DFRL laboratory above me. I was genuinely frightened. But I managed to survive the night. On speaking to our caretaker at about 9 am the following morning, he told me that *'we were burgled again last night'*. Some equipment in the DFRL had been broken.

A.7   Undertaking Original Research at the Daresbury Synchrotron Facility

Early in her research work with me at the RI, Dr Carol Williams agreed to be based at the Science and Engineering Research Council centre in Daresbury to pursue work in which, initially, Richard Catlow (see Section 12.13.2) and our French postdoctoral worker Eric Dooryhee participated. With our sponsors from Unilever and with the ingenuity of Dr Neville Greaves, a Daresbury-based physicist, we were able to combine high-resolution X-ray powder diffraction and X-ray absorption spectroscopy (EXAFS) (see Figure A.10).

To the non-scientist, the key point to note about this work was that, ahead of other experimentalists, we were able to record, in parallel, local (short-range) atomic order and global (long-range) atomic order. In a later study[3] (see Couves et al., Figure A.11 below) that we also carried out at Daresbury alongside Neville Greaves, we were able to trace the conversion of a precursor mineral, known as aurichalcite ($Cu_{5-x}Zn_x(OH)_6(CO_3)_2$), to its active state as a Cu:Zn efficient catalyst for the water-gas shift reaction ($CO_2+H_2 \rightarrow CO+H_2O$). The abstract of this paper, published in *Nature* in 1991, is reproduced in Figure A.11. It was the first-ever study among catalyst experts that combined the merits of X-ray diffraction (for long-range order) and EXAFS (for short-range order).

When, in due course, it became possible to prepare large-pore mesoporous silicas, with diameters that could be as large as 10 Å or more, it became possible to heterogenize bulky organometallic compounds in these nanoporous receptacles.

Faraday Discuss. Chem. Soc., 1990, **89**, 119-136

Structural Studies of High-area Zeolitic Adsorbents and Catalysts by a Combination of High-resolution X-Ray Powder Diffraction and X-Ray Absorption Spectroscopy

Eric Dooryhee† and G. Neville Greaves

S.E.R.C. Daresbury Laboratory, Warrington WA4 4AD

Andrew T. Steel, Rodney P. Townsend and Stuart W. Carr

Unilever Research, Port Sunlight Laboratory, Quarry Road East, Bebbington, Merseyside L63 3JW

John M. Thomas* and C. Richard A. Catlow

Davy Faraday Research Laboratory, The Royal Institution, 21 Albermarle Street, London W1X 4BS

We have characterized at high temperature a model uniform heterogeneous catalyst for the oligomerization of hydrocarbons (a nickel-exchanged zeolite of initial composition $Na_{59}Al_{59}Si_{133}O_{384} \cdot xH_2O$ treated with an aqueous solution of $NiCl_2$ so as to yield a homogeneous distribution of Ni, with $Si/Ni = 7$) by recording the extended X-ray absorption fine structure (EXAFS) above the Ni edge and also its high-resolution diffraction pattern (at $\lambda = 1.5486$ Å). We have obtained unique insights into the microenvironment of the Ni^{2+} ions in the as-prepared and the dehydrated as well as the reduced state of the catalyst; in particular, values of atomic coordinates, site-occupancy and bond lengths, have been obtained.

Figure A.10 *The work of Dooryhee et al. as announced in* Faraday Discuss., **1990**, *89*, 119. *(Courtesy John Meurig Thomas)*

A definitive study that my team reported in *Nature* in 1995 illustrates the point. Figure A.12 (prepared by my collaborator, Professor Avelino Corma in Valencia, Spain) illustrates the nature of the mesoporosity of this silica host.

One of Professor Corma's postdoctoral workers, Fernando Rey, was sent to work in my team at the RI, and he was quite expert in preparing high-quality, open, mesoporous silicas like the image (by computer graphics) of it shown in Figure A.12. Fortunately, one other of my postdoctoral workers, Dr (now

Tracing the conversion of aurichalcite to a copper catalyst by combined X-ray absorption and diffraction

John W. Couves*, John Meurig Thomas*‡, David Waller*, Richard H. Jones*, Andrew J. Dent†, Gareth E. Derbyshire† & G. Neville Greaves†‡

* Davy Faraday Research Laboratory, The Royal Institution of Great Britain, 21 Albermarle Street, London W1X 4BS, UK
† The SERC Daresbury Laboratory, Daresbury, Warrington, Cheshire WA4 4AD, UK

EVER since X-ray sources first became available, the merit of deploying diffraction and absorption spectroscopic studies simultaneously has been acknowledged[1]. Information on oxidation states and local (~ 6-Å radius) atomic environments is now obtained routinely from X-ray absorption measurements using synchrotron sources[2-4]. Synchrotron radiation is also used commonly for high-resolution powder diffraction crystallography. We report here an instrumental arrangement that has allowed us to extract quantitative short- and long-range structural information on samples undergoing chemical change by measuring X-ray absorption spectra and X-ray diffraction patterns *in situ* and within a few seconds of one another, using a synchrotron X-ray source. To illustrate the combination of these techniques, we have followed the structural and chemical changes that occur within the layered mineral aurichalcite ($Cu_{5-x}Zn_x(OH)_6(CO_3)_2$) when heated in dry air to $\sim 450\,°C$. Despite marked changes in crystallinity, the local environment and electronic state of the Cu^{2+} ions remain unchanged, even when at $\sim 450\,°C$ the material is converted to a mixture of CuO and ZnO. Heating this mixture in H_2/N_2 produces an active catalyst for the water–gas shift reaction ($CO_2 + H_2 \rightarrow CO + H_2O$), which our studies show to consist of small particles of copper metal (with some zinc incorporated) supported on ZnO.

Figure A.11 *The work of Couves et al. as announced in Nature,* **1991,** *345.*[4]

(Courtesy John Meurig Thomas)

Figure A.12 *Computer graphic of a typical mesoporous silica.*
(Courtesy Professor Avelino Corma)

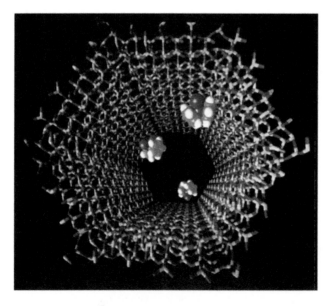

Figure A.13 *Computer-generated illustration of the accommodation (diffusion/adsorption) of molecules of titanocene dichloride inside a pore (30-Å diameter) of siliceous MCM-41. For simplicity, none of the pendant≡Si-OH (silanol) groups, which make it possible to graft organometallic moieties inside the mesoporous host, is shown. Silicon (yellow); oxygen (red); carbon (green); hydrogen (white); chlorine (purple); titanium (blue).*
(Courtesy John Meurig Thomas)

Professor) Gopinat Sankar, was based in Daresbury and became an expert practioner there. The fourth member of our team, Dr Thomas Maschmeyer, a brilliant Australian postdoctoral worker, who joined me at the RI from the University of Sydney, possessed the know-how to prepare and to incorporate bulky organometallic materials into these mesoporous silicas, thereby designing new catalysts Figure A.13.

The paper by Maschmeyer, Rey, Sankar, and Thomas,[4] *'Heterogeneous Catalysts Obtained by Grafting Metallocene Complexes onto Mesoporous Silica'*, has since been cited more than thirteen-hundred times and has become a classic. The abstract reads as follows: *'The synthesis of mesoporous siliceous solids with large-diameter channel apertures (25–100 Å) has greatly expanded the capabilities of heterogeneous catalysis. The large apertures in such mesoporous silicas can, for example, be modified by framework substitution to create highly selective catalysts. Here we show that direct grafting of an organometallic complex onto the inner walls of mesoporous silica MCM-41 generates a shape-selective catalyst with a large concentration of accessible, well spaced and structurally well defined active sites. Specifically, attachment of a titanocene-derived catalyst precursor to the pore walls of MCM-41 produces a catalyst for the epoxidation of cyclohexene and more bulky cyclic alkenes.'*

A.7.1   Single-Site Heterogeneous Catalysts (2012)

Later experimental work in the DFRL ventured into the now burgeoning field of single-site heterogeneous catalysts. Essentially, all the metal-organic framework (open) structures fall into this category. Their special merits are: they are nanoporous, so essentially all the solid is accessible as catalytic centres; they are thermally quite stable; and they are crystalline, so they are amenable to all the techniques of structural characterization and modifications. I wrote *'Design and Application of Single-Site Heterogeneous Catalysis'* [5] as a result, which I was flattered to see was described by Professor Roald Hoffmann as: *'a true marriage of the practical and the fundamental. With the synthetic flair of Humphry Davy and the physical brilliance of his hero Faraday, we are led by the author to a feast of contemporary masterworks of chemical reactivity, prodded by design, into the service of humanity.'*

A.8   Other Pleasures and Privileges of Being Director of the RI

In view of my position at the RI, I was frequently invited to lecture around the world. I was asked to join the Japanese Nobel Laureate Professor Kenishi Fukui to participate in a Tokyo symposium in November 1986 to celebrate the Anniversary of the

Reign of the Emperor of Japan. At very regular intervals thereafter, I was called upon to lecture either on *'The Genius of Michael Faraday'* or on a *'Tale of Contrasting Geniuses: Davy and Faraday'*. Talks were given at more than 40 different universities around the world, from New Zealand to Shanghai to Stanford to Moscow.

Because of the reputation of the RI, the lectures attracted large audiences.

A.8.1   Inviting Christmas Lecturers

The importance of these lectures for the reputation of the RI, as has been discussed in previous chapters, is considerable.

I always made a point of selecting speakers who were blessed with the gift of the spoken word. This was the basis that led me to select Richard Dawkins, Malcolm Longair, and Sir Gareth Roberts. One particular series sticks in my mind. For the 1988 series, I decided to invite one of the most impressive scientists and lecturers in the US—Professor George C. Pimentel, of Berkeley, California, the discoverer of the chemical laser. Earlier that year, he and a mutual friend (Professor Mansel Davies of Aberystwyth) visited me one Friday afternoon at the RI. I gave them a thorough tour of the laboratory and the Faraday museum. Later, at dinner I asked Pimentel if he would be willing to present the 1988 Christmas Lectures on the title *'The Candle Revisited'*. He jumped at the suggestion.

Within a few weeks, he had sent me detailed notes of the contents of each of his lectures. He even outlined some of the demonstrations that he would show. Alas, Pimentel became terminally ill later that year and his death in 1989 was universally mourned. We made him, posthumously, an Honorary Member of the RI.

In his place, Professor Charles A. Taylor stepped in, one of our Visiting Professors (from Cardiff University). Sponsored by the Shell Company, he gave a magnificent series on *'The Physics of Sound'*. Taylor was himself capable of 'extracting sound'— as he put it—from twenty-two different musical instruments! One of his special demonstrations was to drop from his hand a collection of some dozen or so differently shaped blocks of wood. As they fell to the floor, the tune *'twinkle, twinkle, little star'* could be clearly discerned. In his memorable series (which he later gave in Tokyo when he accompanied me to take the RI Christmas Lectures in Japan in 1990), one little girl, aged about twelve, blurted out a question: *'Sir, how far does sound travel?'*

It was always a particular pleasure for the Director to phone up the Christmas Lecturer that it had been decided to invite. In his book *'Brief Candle in the Dark: My Life in Science'*, Richard Dawkins recalls well the exchange that passed between us, while George Porter was proud of the fact that he had invited Carl Sagan to his Christmas Lecture series on *'The Planets'*.

A.9   The Royal Family and the RI

Continuing their long tradition of participating occasionally and supporting generally many of the activities of the RI, present members of the Royal Family are no exception. Earlier we saw (Chapter 1) Prince Albert and his two sons attending a Faraday Christmas Lecture in 1856. In 1896, when Dr Ludwig Mond made his generous bequest to the RI, which resulted in the formation of the DFRL, the Prince of Wales was there to inaugurate it (see Figure A.14).

Shortly after it was announced, in September 1987, that I would (with Professor David Phillips) be giving the RI Christmas Lecture that year, I was invited to lunch at Buckingham Palace. It was interesting to be told by HRH The Queen that, as a family, part of their Christmas television viewing entailed watching the RI Christmas Lectures. To my delight also attending that lunch was my fellow countryman Ian Woosnam, who a short while earlier had won the US Open Golf Tournament.

In 1950, the present Queen, then HRH Princess Elizabeth, was made an Honorary Member of the RI. In my day as Director, the RI was honoured by her presence, and that of HRH The Duke of Edinburgh, when the London

Figure A. 14 *The Prince of Wales inaugurates the Davy–Faraday Research Laboratory (DFRL) in 1896.*

(Author unknown)

Figure A.15 *Sir John Meurig Thomas with HRH Duke of Kent. The photograph was taken at the University of Surrey when Sir John received an Honorary Doctorate from its Chancellor, HRH The Duke of Kent.*

Symphony Orchestra celebrated one of its major milestones in our lecture theatre.

In George Porter's day as Director, he and the Professor of Physics (Harry Messel) at the University of Sydney, established a scheme whereby five scholars of high-school age from Australia (competitively chosen) were invited to spend two days in London, with one of the events there entailing being taken by the Director of the RI to Buckingham Palace to be presented to HRH The Duke of Edinburgh, or, if for logistical reasons that were not possible, a member of the Royal Family coming to the RI to award prizes. At such events, I had to describe exactly what each Australian and British student had done, scientifically, to earn their reward. On one occasion, I went with the students to Buckingham Palace, while on other occasions HRH Princess Anne and later HRH Prince Michael of Kent came to the RI to present the prizes. While talking to the Australian contingent, most of whom came from Sydney, Princess Anne made the amusing remark, that originated from her grandmother (HRH The Queen Mother) that the Sydney Opera House looked like *'three nuns in a rugby scrum!'*

A.9.1   The Duke of Kent

Long before I took up my post at the RI, and still to this day, HRH The Duke of Kent serves as its President. He frequently attended RI Discourses, but in numerous other ways he gave real assistance in furthering its cause, as, for example, in facilitating significant help provided by the Worshipful Company of Clothworkers. He also made many speeches that were much appreciated by members of the RI.

On the occasion of my departure from the RI on 27 June 1991, Sir Paul Osmond, the Chairman of Council, organized an evening party. He informed members and their guests that they were gathered to celebrate an inevitably bittersweet occasion: *'As we know, we are here to take a formal farewell to the Director and Lady Thomas; fortunately not a final farewell since he will have responsibilities that he will continue to exercise here and we shall see him from time to time.'* He then called upon the President, HRH The Duke of Kent, to make formal speech in which the Duke correctly identified Faraday as something of a keynote and inspiration for me during my time at the RI.

HRH also drew attention to the fact *'that the appeal of Faraday had acted as something of a magnet in deciding John Thomas to take up the post of Director'.* Indeed it had. Not only did I hold Faraday in awe from the time I heard Miss James talk about him when I was a schoolboy, there were other reasons why I revered Faraday. In particular, as a young lecturer in the University College of Wales, Bangor, I had set about systematically to explore the chemical consequences of imperfections in solids. About ten years after making that decision, which led to considerable success, I discovered that Faraday had already remarked in his diary that if a knife was used to scratch the surface of a salt-hydrate, like gypsum ($CaSO_4 \cdot 2H_2O$) or sodium carbonate decahydrate ($Na_2CO_3 \cdot 10H_2O$), there would be accelerated efflorescence (loss of water) at the damaged surfaces.

When I was elected Fellow of the Royal Society in 1977, members of the academic staff presented me with a signed copy of one of Faraday's early papers (Figure A.16): his 1825 discovery of benzene and his determination of its composition and properties.

It was also an opportunity for me to recognise my wife's contribution to my tenure and activities as Director. I cannot thank Margaret enough for the elegant and effective way in which she undertook these tasks.

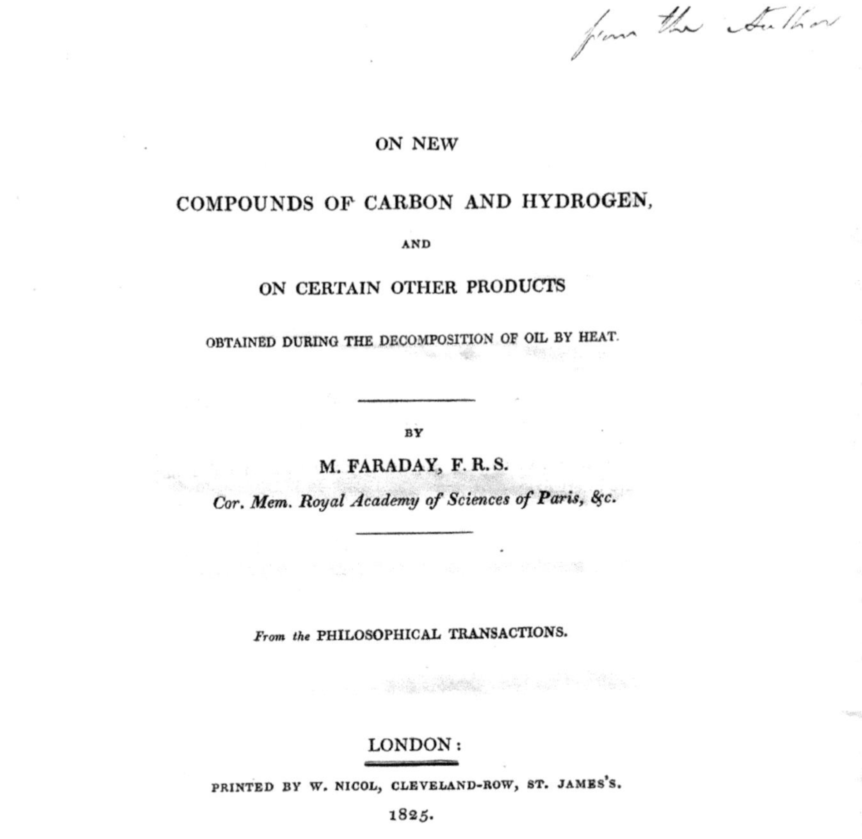

from the author

ON NEW

COMPOUNDS OF CARBON AND HYDROGEN,

AND

ON CERTAIN OTHER PRODUCTS

OBTAINED DURING THE DECOMPOSITION OF OIL BY HEAT.

BY

M. FARADAY, F. R. S.

Cor. Mem. Royal Academy of Sciences of Paris, &c.

From the PHILOSOPHICAL TRANSACTIONS.

LONDON:

PRINTED BY W. NICOL, CLEVELAND-ROW, ST. JAMES'S.

1825.

Figure A.16 *Faraday's 1825 paper; a reprint signed by him.*

The President, Members, and their guests then wandered through the rooms of the RI viewing the exhibition, listening to the quartet, taking refreshments, and meeting their friends. This party marked the end of yet another era in the life of The Royal Institution.

It was Miss Judith Wright, MBE, who first entered the RI as Sir George Porter's personal assistant when she, like him, left Sheffield University to come to Albemarle Street, who contributed greatly to the evening's organisation. She was an outstanding and crucial employee of the Royal Institution for the thirty years she worked there.

Figure A.17 *The bathroom furniture in the Director's flat, with the signature of Faraday.* *(Courtesy the RI)*

It was a perennial thrill, and a profound inspiration, to work in the same study, and to sit on the same chair, as those used by the exemplary Faraday. Most days for five years, this thought would cross my mind. But what influenced me most as I came to the end of each working day was to see Faraday's bathroom furniture, with his signature (see Figure A.17). As I gazed at it, I somehow felt, knowing how prodigiously hard Faraday worked, that I had not worked hard enough to deserve a night's sleep.

A.10 My Views on the Present Status of the RI

For more than two centuries the RI was renowned for: (a) original scientific research; (b) the dissemination of scientific knowledge among the general population; and (c) for the retention of the historicity of its ancient buildings, as the home of some of the greatest scientific discoveries in the world. A subsidiary feature, ever since they were started in 1825, was the maintenance of a high standard and genuinely popular series of RI Christmas Lectures. As will be apparent, it has been a huge privilege for me to serve as Director and I am proud of the work we carried out there as a team, in research and in continuing the tradition of popularizing science to a wide audience.

It is no secret that from the early 1990s onward many significant changes were introduced at the RI. But one of the most important qualities is that it still maintains an outstanding reputation insofar as popular lectures are concerned. The quality of speakers, and their topics for Discourse, are as high as they have ever been, and are shown on the RI channel.

Insofar as the Christmas Lectures are concerned, the situation is less satisfactory. For more than one-hundred-and-sixty years (except during World War II), ever since Faraday started them, there were six lecture-demonstrations given in each series of Christmas Lectures. Now, each series is reduced to three or four lectures. This means that the detailed science that can be conveyed to a 'juvenile' audience, as was done by Faraday and his successors, is limited.

Moving to the sphere of original research, the situation is even less satisfactory I believe that this activity is a mere shadow of what the DFRL did in the period from the 1900s, when it was founded, up until *ca.* 1995. The hundreds of major scientific papers per year that were published from the DFRL in days of yore have effectively ceased. The RI cannot any longer—at least for the present and immediate future—compete in the quest for securing new knowledge on a world scale with other research centres. It is no longer the cradle of discovery and innovation. This saddens me.

The general workshop no longer exists. Offices which hitherto were occupied by scientists are now rented out for other purposes. In addition, the array of expert technicians that was employed by the RI has almost vanished. (No longer is the Bill Coates character around, who could assist the scientific staff in creating new instruments.)

Unless major changes of policy are made in the near future, I believe that there is little prospect now of regenerating the DFRL as a world-class stand alone centre of scientific research.

REFERENCES

1. L. N. Johnson and G. A. Petsko, *Nature*, **1999**, *399*, 26.
2. L. N. Johnson and G. A. Petsko, *Trends Biochem. Sci.*, **1999**, *24*, 287.
3. J. W. Couves, J. M. Thomas, and G. E. Greaves et al., *Nature*, **1991**, *354*, 465.
4. T. Maschmeyer, G. Sankar, F. Rey, and J. M. Thomas, *Nature*, **1995**, *378*, 160.
5. J. M. Thomas, *'Design and Application of Single-Site Heterogeneous Catalysts: Applications of Clean Technology, Green Chemistry, and Sustainability'*, Imperial College Press, **2012**.

Index: *Albemarle Street*

Note: Tables and figures are indicated by an italic *t* and *f* following the page number.